高等职业教育"十三五"规划教材

生物技术专业英语

汤卫华 陈 珊 主编

中国轻工业出版社

图书在版编目（CIP）数据

生物技术专业英语/汤卫华，陈珊主编. —北京：
中国轻工业出版社，2017.6
高等职业教育"十三五"规划教材
ISBN 978-7-5184-1360-7

Ⅰ.①生… Ⅱ.①汤… ②陈… Ⅲ.①生物工程-英语-高等职业教育-教材 Ⅳ.①Q81

中国版本图书馆 CIP 数据核字（2017）第 074035 号

责任编辑：江 娟 贺 娜

策划编辑：江 娟　　责任终审：张乃东　　封面设计：锋尚设计
版式设计：宋振全　　责任校对：晋 洁　　责任监印：张 可

出版发行：中国轻工业出版社（北京鲁谷东街5号，邮编：100040）
印　　刷：三河市万龙印装有限公司
经　　销：各地新华书店
版　　次：2017年6月第1版第1次印刷
开　　本：720×1000 1/16 印张：16.5
字　　数：321千字
书　　号：ISBN 978-7-5184-1360-7 定价：40.00元
邮购电话：010-85119873
发行电话：010-85119832 010-85119912
网　　址：http://www.chlip.com.cn
Email：club@chlip.com.cn
版权所有 侵权必究
如发现图书残缺请直接与我社邮购联系调换
KG1486—151279

编写人员名单

主　编　汤卫华　陈　珊
副主编　张　乐　龙　尾　吕春晖　闫雪冰
参　编　苑　鹏　高芦宝

前　　言

　　《普通高等学校高等职业教育专科（专业）目录（2015年）》是高等职业教育的基本指导性文件，是高校设置与调整高职专业、实施人才培养、组织招生、指导就业的基本依据。《生物技术专业英语》以该目录的专业划分和调整为基础，以高职教育专业理论"必需、够用"为原则，在深度上考虑到高职学生的掌握程度，以浅显易懂为主。目前，适用于高职高专层次的同类教材非常少，而本科教材较多。本教材的优势是遵循高职学生的认知程度，在难度上降低，增强其适用性。另外，80%的参考资料选自5年内出版的书籍或期刊，以提升课本的时效性和前沿性。

　　本书共分为7个单元，具体编写分工如下：单元1为化学基础知识，由高芦宝编写；单元2为生物化学，由张乐编写；单元3为微生物，由龙尾编写；单元4为发酵工程，由陈珊编写；单元5为酶工程，由汤卫华和闫雪冰编写；单元6为药物制剂，由吕春晖编写；单元7为GMP，由苑鹏编写。每章包括阅读材料、词汇、参考译文及拓展阅读。内容覆盖食品生物技术和药品生物技术的基本领域，突出了行业发展方向。

　　《生物技术专业英语》可作为食品生物技术、药品生物技术等高职高专专业和应用型本科的专业英语教材或参考用书。

目　　录

Unit 1　Chemical Basic Knowledge ································· 1
　Chapter 1　Atoms and Molecules ································· 1
　　1. Elements and compounds ································· 1
　　2. Chemical reactions and physical changes ················ 1
　　3. Atoms and molecules ································· 3
　Chapter 2　Organic Compounds ································· 10
　　1. What is an organic compound? ························· 10
　　2. Kinds of organic compounds ··························· 11
　　3. The saturated hydrocarbons, or alkanes ················ 12
　　4. Structure ··· 13

Unit 2　Biochemistry ·· 18
　Chapter 1　Chemical Elements of Life ···························· 18
　　1. The origin of life ·· 18
　　2. The chemical elements of life ·························· 20
　Chapter 2　Biomolecules: Carbohydrates, Lipids and Protein ··· 25
　　1. Many important macromolecules are polymers ··········· 25
　　2. Carbohydrates ·· 26
　　3. Lipids ·· 28
　　4. Protein ··· 29
　Chapter 3　Glycolysis ··· 36
　　1. Glycolysis pathway in many organisms ·················· 37
　　2. The citric acid cycle ····································· 39
　Chapter 4　Diabetes and Human Health ··························· 46
　　1. What is diabetes mellitus? ······························· 46
　　2. What causes diabetes? ····································· 46
　　3. Diabetes diagnosis ·· 47
　　4. Regulation of glucose transporter translocator in health and diabetes ··· 47

Unit 3　Microbiology ··· 57
　Chapter 1　Introduction of Microbiology ························· 57
　　1. The discovery of microorganisms ······················· 57

  2. The relationship between microorganisms and diseases ⋯⋯⋯⋯ 58

  3. The development of techniques for studying microbial pathogens ⋯⋯⋯⋯ 60

  4. Members of the microbial world ⋯⋯⋯⋯⋯⋯⋯⋯⋯⋯⋯⋯⋯⋯⋯⋯ 62

 Chapter 2 Microorganism and Culture Media ⋯⋯⋯⋯⋯⋯⋯⋯⋯⋯⋯⋯ 68

  1. The common nutrient requirements ⋯⋯⋯⋯⋯⋯⋯⋯⋯⋯⋯⋯⋯⋯⋯ 68

  2. Nutritional types of microorganisms ⋯⋯⋯⋯⋯⋯⋯⋯⋯⋯⋯⋯⋯⋯⋯ 69

  3. Uptake of nutrients by the cell ⋯⋯⋯⋯⋯⋯⋯⋯⋯⋯⋯⋯⋯⋯⋯⋯⋯ 71

  4. Culture media ⋯⋯⋯⋯⋯⋯⋯⋯⋯⋯⋯⋯⋯⋯⋯⋯⋯⋯⋯⋯⋯⋯⋯⋯ 71

 Chapter 3 Microbial Growth ⋯⋯⋯⋯⋯⋯⋯⋯⋯⋯⋯⋯⋯⋯⋯⋯⋯⋯⋯ 76

  1. The growth curve ⋯⋯⋯⋯⋯⋯⋯⋯⋯⋯⋯⋯⋯⋯⋯⋯⋯⋯⋯⋯⋯⋯ 76

  2. The continuous culture of microorganisms ⋯⋯⋯⋯⋯⋯⋯⋯⋯⋯⋯⋯ 77

 Chapter 4 Recent Developments in the Production of Valuable

      Microbial Products ⋯⋯⋯⋯⋯⋯⋯⋯⋯⋯⋯⋯⋯⋯⋯⋯⋯⋯ 83

Unit 4 Fermentation Engineering ⋯⋯⋯⋯⋯⋯⋯⋯⋯⋯⋯⋯⋯⋯⋯⋯ 92

 Chapter 1 Fermentation Microbiology and Biotechnology ⋯⋯⋯⋯⋯⋯⋯ 92

  1. The nature of fermentation ⋯⋯⋯⋯⋯⋯⋯⋯⋯⋯⋯⋯⋯⋯⋯⋯⋯⋯⋯ 92

  2. Chemical and biological engineering ⋯⋯⋯⋯⋯⋯⋯⋯⋯⋯⋯⋯⋯⋯⋯ 97

 Chapter 2 Strain Screen ⋯⋯⋯⋯⋯⋯⋯⋯⋯⋯⋯⋯⋯⋯⋯⋯⋯⋯⋯⋯ 108

  1. Selection and screening ⋯⋯⋯⋯⋯⋯⋯⋯⋯⋯⋯⋯⋯⋯⋯⋯⋯⋯⋯⋯ 108

  2. Mutagenesis ⋯⋯⋯⋯⋯⋯⋯⋯⋯⋯⋯⋯⋯⋯⋯⋯⋯⋯⋯⋯⋯⋯⋯⋯ 110

 Chapter 3 Fermentation Medium ⋯⋯⋯⋯⋯⋯⋯⋯⋯⋯⋯⋯⋯⋯⋯⋯⋯ 116

  1. Fermentation medium ⋯⋯⋯⋯⋯⋯⋯⋯⋯⋯⋯⋯⋯⋯⋯⋯⋯⋯⋯⋯ 116

  2. Growth medium ⋯⋯⋯⋯⋯⋯⋯⋯⋯⋯⋯⋯⋯⋯⋯⋯⋯⋯⋯⋯⋯⋯ 117

 Chapter 4 Fermenters ⋯⋯⋯⋯⋯⋯⋯⋯⋯⋯⋯⋯⋯⋯⋯⋯⋯⋯⋯⋯⋯ 122

  1. Fermenter ⋯⋯⋯⋯⋯⋯⋯⋯⋯⋯⋯⋯⋯⋯⋯⋯⋯⋯⋯⋯⋯⋯⋯⋯ 122

  2. Other fermenter design and ancillary equipment ⋯⋯⋯⋯⋯⋯⋯⋯⋯ 123

 Chapter 5 Production of Antibiotics ⋯⋯⋯⋯⋯⋯⋯⋯⋯⋯⋯⋯⋯⋯⋯⋯ 130

  1. Antibiotics ⋯⋯⋯⋯⋯⋯⋯⋯⋯⋯⋯⋯⋯⋯⋯⋯⋯⋯⋯⋯⋯⋯⋯⋯ 130

  2. Penicillins ⋯⋯⋯⋯⋯⋯⋯⋯⋯⋯⋯⋯⋯⋯⋯⋯⋯⋯⋯⋯⋯⋯⋯⋯ 131

Unit 5 Enzyme ⋯⋯⋯⋯⋯⋯⋯⋯⋯⋯⋯⋯⋯⋯⋯⋯⋯⋯⋯⋯⋯⋯⋯⋯ 136

 Chapter 1 The Biological Catalysts of Life ⋯⋯⋯⋯⋯⋯⋯⋯⋯⋯⋯⋯⋯ 136

  1. Introduction ⋯⋯⋯⋯⋯⋯⋯⋯⋯⋯⋯⋯⋯⋯⋯⋯⋯⋯⋯⋯⋯⋯⋯ 136

  2. Classification and nomenclature of enzyme ⋯⋯⋯⋯⋯⋯⋯⋯⋯⋯⋯ 138

  3. Enzyme applications ⋯⋯⋯⋯⋯⋯⋯⋯⋯⋯⋯⋯⋯⋯⋯⋯⋯⋯⋯⋯ 140

 Chapter 2 Enzyme Production ⋯⋯⋯⋯⋯⋯⋯⋯⋯⋯⋯⋯⋯⋯⋯⋯⋯⋯ 146

1. Summary ……………………………………………………………………146
  2. Enzyme sources ……………………………………………………………147
  3. Strain development ………………………………………………………149
  4. Growth requirements of microorganisms ……………………………151
  5. Fermentation ………………………………………………………………152
  6. Regulations during enzyme production………………………………155
 Chapter 3　Immobilized Enzyme Technology ……………………………… 162
  1. Immobilization techniques ……………………………………………162
  2. Benefits and characteristics …………………………………………170

Unit 6　Production of Dosage Forms ……………………………………… 179
 Chapter 1　Introduction to Drugs ……………………………………………179
  1. What is a drug …………………………………………………………179
  2. Drug bioavailability …………………………………………………182
 Chapter 2　Oral Preparation ………………………………………………… 188
  1. Dosage forms application ……………………………………………188
  2. Absorption ………………………………………………………………191
 Chapter 3　Parenteral Administration ……………………………………… 197
  1. Introduction ……………………………………………………………197
  2. Routes of parenteral administration …………………………………198
  3. The preparation of parenteral products ……………………………200
 Chapter 4　Drug for External Use ………………………………………… 210
  1. Epicutaneous route ……………………………………………………210
  2. Ocular, otic, and nasal routes…………………………………………213

Unit 7　Good Manufacturing Practice for Drugs ……………………… 218
 Chapter 1　Process Validation: General Principles and Practices……………… 218
  1. General considerations for process validation ………………………218
  2. Stage 1——process design ……………………………………………219
  3. Stage 2——process qualification ……………………………………223
  4. Stage 3——continued process verification …………………………230
 Chapter 2　Non-Penicillin β-Lactam Drugs: A CGMP Framework for
        Preventing Cross-Contamination………………………………… 243
  1. Regulatory framework …………………………………………………243
  2. β-lactam antibiotics ……………………………………………………244
  3. β-lactamase inhibitors …………………………………………………246
  4. β-lactam intermediates and derivatives ……………………………247

Unit 1　Chemical Basic Knowledge　化学基础知识

Chapter 1　Atoms and Molecules　分子和原子

1. **Elements** and compounds 　　There are two types of pure substance: elements and compounds. ● Elements are substances that cannot be **chemically** broken down into simpler substances. ● Compounds are pure substances made from two, or more, elements chemically combined together. 　　Elements are the "building blocks" from which the Universe is constructed. There are over a hundred known elements, but the vast majority of the Universe consists of just two. **Hydrogen** (92%) and helium (7%) make up most of the mass of the Universe, with all the other elements contributing only 1% to the total. The concentration, or "coming together", of certain of these elements to make the Earth is of great interest and significance. There are 94 elements found naturally on Earth altogether. Two elements, **silicon** and **oxygen**, which are bound together in silicate rocks. Only certain of the elements are able to form the complex compounds that are found in living things. For example, the human body contains 65% oxygen, 18% carbon, 10% hydrogen, 3% **nitrogen**, 2% **calcium** and 2% of other elements. 2. Chemical reactions and physical changes 　　Substances can mix in a variety of ways, and	**New Words and Expressions** element ['elɪmənt] *n.* 元素；要素 chemically ['kemɪklɪ] *adv.* 用化学，以化学方法 hydrogen ['haɪdrədʒən] *n.* 氢 silicon ['sɪlɪkən] *n.* 硅；硅元素 oxygen ['ɒksɪdʒən] *n.* 氧气，氧 nitrogen ['naɪtrədʒən] *n.* 氮 calcium ['kælsiəm] *n.* 钙

they can also react chemically with each other. In a reaction, one substance can be transformed (changed) into another. **Copper carbonate** is a green solid, but on heating it is changed into a black powder (Figure 1-1). Closer investigation shows that the gas **carbon dioxide** is also produced. This type of chemical reaction, where a compound breaks down to form two or more substances, is known as **decomposition.**

copper carbonate
[ˈkɔpə ˈkɑːbəneit]
n. 碳酸铜
carbon dioxide
[ˈkɑːbən daɪˈɔksaɪd]
n. 二氧化碳
decomposition
[diːkɒmpəˈzɪʃn]
n. 分解，腐烂；变质

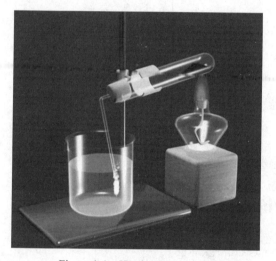

Figure 1-1 Heating copper carbonate

Decomposition can also be brought about by electricity. Some substances, although they do not conduct electricity when solid, do conduct when they are melted or in solution. In the process of conduction, they are broken down into simpler substances. Thus, lead (II) **bromide**, which is a white powder, can be melted. When a current is passed through molten lead (II) bromide, a **silver-grey** metal (lead) and a brown vapour (bromine) are formed. Neither of these products can be split into any simpler substances.

bromide [ˈbrəʊmaɪd]
n. 溴化物
silver-grey [ˈsɪlvəˈgrei]
n. 银灰色

The opposite type of reaction, where the substance is formed by the combination of two or more other substances, is known as **synthesis**. For example, if a pieced of burning magnesium is plunged into a gas jar of oxygen, the intensity (brightness) of the brilliant white flame increase. When the reaction has burnt out, a white ash remains (Figure 1-2). The ash has totally different properties form the original silver-grey metal strip and colorless gas we started with. A new compound, **magnesium oxide,** has been formed from magnesium and oxygen.

synthesis ['sɪnθɪsɪs]
n. 合成；综合体

magnesium oxide
[mæg'niːziːəm 'ɔksaid]
n. 氧化镁

Figure 1-2   Burning magnesium produces a brilliant white flame

Although many other reactions are not as spectacular as this, the burning of magnesium shows the general features of chemical reactions.

3. **Atoms** and molecules

The molecular structure hypothesis-that a molecule is a collection of atoms linked by a network of **bonds** was forged in the crucible of nineteenth century experimental chemistry. It has continued to serve as the principal means of ordering and classifying the observations of chemistry. The difficulty with this hypothesis was that it was not related directly to **quantum**

atom ['ætəm]
n. 原子；原子能

bond [bɒnd]
n. 化学键

quantum ['kwɒntəm]
n. 量子；定量，总量

mechanics, the physics which governs the motions of the **nuclei** and electrons that make up the atoms and the bonds. Indeed, there was, and with some there still is, a prevailing opinion that these fundamental concepts, while unquestionably useful, were beyond theoretical definition. We have in chemistry an understanding based on a classification scheme that is both powerful and at the same time, because of its empirical nature, limited.

  Richard Feynman and Julian Schwinger have given us a reformulation of physics that enables one to pose and answer the questions "what is an atom in a molecule and how does one predict its properties?" These questions were posed in my laboratory where it was demonstrated that this new formulation of physics, when applied to the observed **topology** of the distribution of **electronic charge** in real space, yields a unique partitioning of some total system into a set of bounded spatial regions. The form and properties of the groups so defined faithfully recover the characteristics ascribed to the atoms and functional groups of chemistry. By establishing this association, the molecular structure hypothesis is freed from its empirical constraints and the full predictive power of quantum mechanics can be incorporated into the resulting theory-a theory of atoms in molecules and crystals.

  The theory recovers the central operational concepts of the molecular structure hypothesis, that of a functional grouping of atoms with an additive and characteristic set of properties, together with a definition of the bonds that link the atoms and impart the structure. Not only does the theory

nuclei ['njuːklɪaɪ]
n. 核心，核子；原子核

topology [təˈpɒlədʒɪ]
n. 拓扑结构
electronic charge [ilek'trɔnik tʃɑːdʒ]
n. 电子电荷

thereby quantify and provide the physical understanding of the existing concepts of chemistry, it makes possible new applications of theory. These new applications will eventually enable one to perform on a computer, in a manner directly paralleling experiment, everything that can now be done in the laboratory, but more quickly and more efficiently, by linking together the functional groups of theory. These applications include the design and synthesis of new molecules and new materials with specific desirable properties.

Matter is composed of atoms. This is a consequence of the manner in which the electrons are distributed throughout space in the attractive field exerted by the nuclei. The nuclei act as point attractors immersed in a cloud of **negative charge**, the **electron density** (r). The electron density describes the manner in which the electronic charge is distributed throughout real space. The electron density is a measurable property and it determines the appearance and form of matter. This is illustrated in the following figures. Figure 1-3 displays the spatial distribution of the electron density in the plane containing the two carbon and four hydrogen nuclei of the **ethane** molecule. The electron density is a maximum at the position of each nucleus and decays rapidly away from these positions. When this diagram is translated into three dimensions, the cloud of negative charge is seen to be most dense at **nuclear** positions and to become more diffuse as one moves away from these centres of attraction, as illustrated in Figure 1-3.

negative charge
['negətiv tʃɑːdʒ]
n. 负电荷
electron density
[ɪ'lektran 'densiti]
n. 电子密度

ethane ['eθiːn]
n. 乙烯

nuclear ['njuːklɪə]
adj. 原子能的；细胞核的；中心的；原子核的

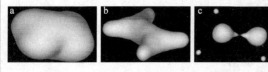

Figure 1-3　Envelopes of the electron density for the ethene molecule

参考译文

1. 元素和化合物

纯物质有两种，元素和化合物。

- 元素不能够通过化学方法分解成为更简单的物质。
- 化合物是由两个或者更多化学元素组合在一起的物质。

元素是构成世界的"基石"。目前已知的元素有一百多种，但是构成宇宙中绝大多数的物质主要有两种，氢（92%）和氦（7%）。这两种元素在宇宙的物质中占到了很高的质量分数，其他元素只占总体质量分数1%的浓度，但这些元素对于世界物质的构成具有极大的意义。在自然界存在着94种元素。硅和氧两种元素共同存在于硅酸盐岩石中。在生物体中只有特定的元素可以形成复杂的化合物，例如，人体包含65%的氧、18%的碳、10%的氢、3%的氮、2%的钙和2%的其他元素。

2. 化学反应和物理变化

物质的混合方式有多种，不同物质之间还可以发生化学反应。在一种反应中，一种物质可通过反应变成另一种新物质。碳酸铜是一个绿色的固体，通过加热可以变成黑色粉末状物质（图1-1）。反应同时也会产生二氧化碳气体。这种类型的化学反应，将一种化合物进行分成两个或两个以上的物质称为分解反应。

图1-1　加热碳酸铜

一些溶解在溶液中的物质尽管不导电，也可以通过电解的方式进行分解反应。在反应过程中，它们被分解成结构更简单的物质。因此，白色粉末状的溴化铅可以被溶解。当溴化铅被电解时，会产生银灰色的金属铅和棕色的溴气。这两种物质不能再进一步分解。

与分解反应相反，由两种或两种以上物质合成为一种物质的反应称为合成。例如，将镁放入氧气罐中燃烧会产生亮白色的火焰。反应完毕后会出现白色的灰状物质（图1-2）。这种物质与银色金属物质和气体物质的性质完全不同，这是一种新的化合物氧化镁，是由镁和氧气合成的。

图1-2 燃烧镁产生的白色火焰

虽然其他一些反应不像此反应这么强烈，但是镁的燃烧反应体现了化学反应的普遍特征。

3. 原子和分子

分子结构的假说，一个分子是由原子以网状结构连接而形成的集合体，这是19世纪进行化学实验得出的结论。这种方法一直成为化学研究的主要手段。这一假设的难点是它与量子力学没有直接的关系，原子核和电子的运动构成了原子及其连接机构。实际上还有一些流行的观点认为这些基本概念已经超出了基础理论的认知。基于专业分类体制，我们对化学领域的研究具备了有效性以及同时性，但是由于经验的缺乏我们的研究还是有限制的。

理查德·费曼物理学和朱利安·施温格提出并回答了一个问题"什么是原子和分子，如何预测它的属性？"在我的实验室里当应用于观察到的拓扑分布的电子电荷在现实空间中产生了一个独特的一些总系统，并划分为一组有限的空间区域时，这个问题被新的公式证明了。功能团的形成方式和特点定义为原子和官能团的化学。通过建立这种联系，分子结构假说从其经验约束和量子力学的理论完整推测可以得出最终的理论——原子和分子以及晶体的理论。

这个理论引出了分子结构假说理论的核心概念：原子功能团的特征以及性质，以及连接原子的结构和定义。这些理论从量化的角度帮助理解了现有的化学理论及概念，它使得新的理论有了应用的空间。这些新的理论在实验室中就可以完成研究，通过应用程序的方式，利用计算机模拟以及实验的方法来观察，能够更有效地发现官能团的研究理论。这些应用程序包括新分子的设计和合成、新材料与特定的属性。

物质是由原子组成的，这是由电子的分布方式以及原子核空间的吸引力决定的。原子核将负电荷电子吸引到电子云周边。电子密度（r）是描述电子的方式在真实空间电荷分布的物理量。电子密度是一个可测量物理量，它决定了原子的外观和性质。图 1-3 显示了在平面上乙烯分子的电子密度空间分布，包含两个碳和四个氢原子核。电子密度的尺寸为从原子核位置起由于衰减的原因远离核的最长距离。当这个图转化为三维空间，负电荷的电子云从核中心的密集区变得更加分散，逐渐远离中心，如图 1-3 所示。

图 1-3　乙烯分子的电子密度

Further Reading

Elements and Compounds

Elements and Compounds All matter is composed of basic substances called elements. An element cannot be broken down into simpler units by chemical reactions; it contains only one kind of atom. An atom is the smallest characteristic unit of an element. A compound is a substance that can be split into two or more elements. Water is a compound because it can be split into its components, hydrogen and oxygen. The **formula** of a compound gives information about the kinds and numbers of atoms that make up each molecule of that compound. A formula contains the symbols of the kinds of atom in each molecule and subscripts that indicate the number of each kind of atom in the molecule. For example, the formula for water, H_2O, indicates that a water molecule contains two hydrogen atoms and one oxygen atom; and a molecule of the glucose, $C_6H_{12}O_6$, contains six carbon atoms, twelve hydrogen atoms, and six oxygen atoms. When carbon unites with oxygen, it forms a colorless, odorless, and tasteless gas called carbon dioxide, which is heavier than air and will extinguish a flame. Carbon dioxide is like nitrogen in many ways, but if it is mixed with limewater, it causes the clear liquid to because milky, while nitrogen dose not. This is the test for carbon dioxide. Carbon dioxide is a source of plant food. Plant has the power to take this gas from the air, combine it with water, and make it into their tissues; in fact, it is from this source that all organic carbon comes. **Mineral** compounds are made of elements such as **sulphur**, phosphorus, iron, potassium, sodium, and calcium. Calcium unites with sulphur and oxygen to form calcium sulphate, and phosphors and oxygen to form calcium phosphate, sodium and potassium unite with oxygen and nitrogen to form sodium to potassium nitrates.

New Words

formula ['fɔːmjʊlə] *adj.* 配方

mineral ['mɪnərəl] *n.* 矿物

sulphur ['sʌlfə] *n.* 硫黄；硫黄色

参考文献

[1] Tada, Masashi. Adamantane compound for organic electroluminescent elements, and organic electroluminescent element. EP2985802. 2016.

[2] Yamauchi, Takuya, Inoue, Katsuya, et al. Nickel manganese compound hydroxide particles and method for producing same. WO/2016/013674. 2016.

[3] Reiserer A. A quantum gate between a flying optical photon and a single trapped atom//A controlled phase gate between a single atom and an optical photon. Springer International Publishing, 2016:237–240.

[4] Abdelwahab N H, Thabet L E. On a moving four-level N-type atom interacting with two-mode cavity field in the presence of the cross-Kerr medium. Modern Physics Letters B, 2016.

Chapter 2 Organic Compounds 有机化合物

1. What is an organic compound?

When you drive up to the pump at some gas stations you are faced with a variety of choices. You can buy "leaded" gas or different forms of "unleaded" gas that have different **octane** numbers. As you filled the tank, you might wonder, "What is 'leaded' gas, and why do they add lead to gas?" Or, "What would I get for my money if I bought premium gas, with a higher octane number?"

You then stop to buy drugs for a sore back that has been bothering you since you helped a friend move into a new apartment. Once again, you are faced with choices. You could buy **aspirin**, which has been used for almost a hundred years. Or **Tylenol**, which contains **acetaminophen**. Or a more modern pain-killer, such as **ibuprofen**. While you are deciding which drug to buy, you might wonder, "What is the difference between these drugs?" and even, "How do they work?"

You then drive to campus, where you sit in a "plastic" chair to eat a sandwich that has been wrapped in "plastic," without worrying about why one of these plastics is flexible while the other is rigid. While you're eating, a friend stops by and starts to tease you about the effect of your diet on the level of **cholesterol** in your blood, which brings up the questions, "What is cholesterol?" and "Why do so many people worry about it?"

Answers to each of these questions fall within the realm of a field known as organic chemistry. For more than 200 years, chemists have divided

New Words and Expressions

octane ['ɑkten]
n. 辛烷

aspirin ['æsprɪn]
n. 阿司匹林（解热镇痛药）
tylenol ['taɪlə,nɔl]
n. 泰诺（酚麻美敏混悬液）
acetaminophen [əsitə'mɪnəfɛn]
n. 对乙酰氨基酚
ibuprofen [,aɪbjuː'prəufen]
n. 布洛芬，异丁苯丙酸

cholesterol [kə'lɛstərɔl]
n. 胆固醇

materials into two categories. Those isolated from plants and animals were classified as organic, while those that trace back to minerals were **inorganic**. At one time, chemists believed that organic compounds were fundamentally different from those that were inorganic because organic compounds contained a vital force that was only found in living systems.

Organic compounds are carbon-based compounds. Organic compounds contain carbon bonds in which at least one carbon atom is covalently linked to an atom of another type (usually hydrogen, oxygen or nitrogen). Most polymers are organic compounds.

2. Kinds of organic compounds

There are natural organic compounds, and synthetic ones. Their structure may be described by using names, and making diagrams.

One way of showing the molecule is by drawing its structural formula. Because molecules can have complicated structures, people have made ways to show them in simple language. One way is to use line diagrams. Each atom is shown by a letter, and connected by a line to each atom with which it is has a **covalent** bond. One line means a single bond, two lines means a double bond and so on.

(1) Natural compounds Natural compounds are compounds made by plants or animals. These could also be made in a lab, but many of these compounds are taken from nature because it is easier and less expensive to do it that way. Common natural compounds are: **amino acids**, proteins, carbohydrates, many antibiotics like Penicillin and Amoxicillin.

inorganic ['ɪnɔr'gænɪk]
adj. 无机的；无生物的

covalent [kəʊ'veɪlənt]
adj. 共价的；共有原子价的

amino acid [ə,miːnəu 'æsid]
n. 氨基酸

(2) Synthetic compounds Synthetic compounds are those made by people. Sometimes, this is done by taking something natural and changing the molecule in a small way, such as making **glycerin** from vegetable oils. Other compounds are synthesized in long, complicated reactions with many steps. Plastics are sometimes mostly natural, and other kinds are manufactured.

3. The saturated **hydrocarbons**, or **alkanes**

Compounds that contain only carbon and hydrogen are known as hydrocarbons. Those that contain as many hydrogen atoms as possible are said to be saturated. The saturated hydrocarbons are also known as alkanes.

The simplest alkane is **methane**: CH_4. The Lewis structure of methane can be generated by combining the four electrons in the **valence** shell of a neutral carbon atom with four hydrogen atoms to form a compound in which the carbon atom shares a total of eight valence electrons with the four hydrogen atoms.

Methane is an example of a general rule that carbon is tetravalent; it forms a total of four bonds in almost all of its compounds. To minimize the repulsion between pairs of electrons in the four CH bonds, the geometry around the carbon atom is tetrahedral.

The alkane that contains three carbon atoms is known as propane, which has the formula C_3H_8. The four carbon alkane is **butane**, with the formula C_4H_{10}.

The names, formulas, and physical properties for a variety of alkanes with the generic formula C_nH_{2n+2}. The boiling points of the alkanes gradually increase with the molecular weight of these

glycerin ['glɪsərɪn]
n. 甘油

hydrocarbon [ˌhaɪdrə'kɑːbən]
n. 碳氢化合物，烃
alkane ['ælkeɪn]
n. 链烷，烷烃

methane ['miːθeɪn]
n. 甲烷，沼气
valence ['veləns]
n. 价；原子价；化合价；效价

butane ['bjuːteɪn]
n. 丁烷

compounds. At room temperature, the lighter alkanes are gases; the midweight alkanes are liquids; and the heavier alkanes are solids, or tars.

In addition to the straight-chain examples considered so far, alkanes also form branched structures. The smallest hydrocarbon in which a branch can occur has four carbon atoms. This compound has the same formula as butane (C_4H_{10}), but a different structure. Compounds with the same formula and different structures are known as **isomers**. When it was first discovered, the branched isomer with the formula C_4H_{10} was therefore given the name **isobutane**.

isomer ['aɪsəmə]
n. 同分异构物；同质异能素
isobutane [ˌaɪsə'bjuːteɪn]
n. 异丁烷

The best way to understand the difference between the structures of butane and isobutene is to compare the ball-and-stick models.

Butane and isobutene are called constitutional isomers because they literally differ in their constitution. One contains two CH_3 groups and two CH_2 groups; the other contains three CH_3 groups and one CH group.

4. Structure

Since a compound is often first discovered in nature instead of being made on purpose in a lab, people may know the compound exists, and even know what it does sometimes, but not know exactly what atoms it is made of and how it is arranged. There are several ways of taking an unknown compound and finding out this structure:

Mass spectrometry;

X-ray diffraction;

Nuclear magnetic resonance spectroscopy;

Infrared spectroscopy.

mass spectrometry
[mæs spek'trɔmitri]
n. 质谱法
X-ray diffraction
['eksrei di'frækʃən]
n. X射线衍射
infrared [ˌɪnfrə'red]
adj. 红外线的

参考译文

1. 什么是有机化合物？

当你开车到加油站会有很多种选择，根据不同的辛烷值你可以选择有铅或者无铅气体。当你充满油后你可能会思考，什么是铅？如果我购买了辛烷值更高的汽油会怎么样？

当你感到背部酸痛的时候你可以购买已经使用一百多年的阿司匹林或者含有乙酰氨基酚的泰诺，也可以使用布洛芬等现代的止痛药。当你在使用这些药物的时候你可能会思考，这些药物有什么区别呢？或者他们治疗的原理是什么？

当你开车去学校时你可以坐在塑料的椅子上吃被塑料膜包裹的三明治，你可能好奇为什么一种塑料柔软而一种塑料硬度很高。当你吃饭的时候，你的朋友会提醒你的饮食会影响体内的胆固醇水平，你可能会提出问题：胆固醇是什么？为什么许多人会担心胆固醇的指标？

这些问题属于有机化学领域的范畴。两百多年来，化学家将材料分为两类，植物和动物归为有机类，其他的矿物质称作无机类。化学家认为有机化合物与无机物的根本区别在于有机物在生命体中可以找到。

有机化合物是碳基化合物，有机化合物含有碳键，其中至少一个碳原子与另一种原子（通常是氢、氧或氮）共价连接。大多数聚合物是有机化合物。

2. 有机化合物种类

有机化合物分为天然的和合成的。它们的结构可以用名称和图表来描述。

显示分子的方法之一是通过它的结构式。因为分子可以有复杂结构，人们用简单的语言去表示分子结构。一种方法是用线条图，每个原子是由一个字母表示，由一条线连接到两个原子并形成一个共价键。一条线表示单键，两条线表示双键等。

（1）天然化合物 天然化合物是由植物或动物制成的化合物。这些也可以在实验室中进行，但许多化合物是来源于大自然的，因为更容易和更便宜。常见的天然化合物有氨基酸、蛋白质、碳水化合物、青霉素和阿莫西林等多种抗生素。

（2）合成化合物 合成化合物是由人工制成的，有时，这是通过采取一些自然和改变分子的小方法，如从植物油中提取甘油。其他化合物的合成需要长时间复杂的反应步骤。塑料大多是天然的，有些也是通过塑料制造产生的。

3. 饱和烃或烷烃

只含有碳和氢的化合物被称为碳氢化合物。那些包含尽可能多的氢原子被认为是饱和的。饱和烃也称为烷烃。

最简单的烷烃是甲烷。甲烷的路易斯结构，中性碳原子的四个电子与四个氢原子结合，形成碳原子与四个氢原子共享 8 个价电子的化合物来产生的。

甲烷是碳四价规则的一个例子，它在几乎所有的化合物中总共形成了四个键。为了尽量减少在四个 C—H 键的电子对之间的排斥力，周围碳原子的几何形成是四面体。

包含三个碳原子的烷烃被称为丙烷，化学式为 C_3H_8。四碳烷烃是丁烷，化学式为 C_4H_{10}。

各种烷烃具有通式为 C_nH_{2n+2}。烷烃的沸点随着这些化合物分子量增加而逐渐增加。在室温下，轻烷烃是气体；中等重量的烷烃是液体；较重的烷烃是固体或焦油。

到目前为止，烷烃除了具有直链结构外，烷烃也形成支链结构。具有分支的最小碳氢化合物有四个碳原子。该支链化合物具有跟丁烷（C_4H_{10}）一样的分子式，但是具有不同的结构。化具有相同分子式和不同结构的化合物称为同分异构体。首次被发现分子式为 C_4H_{10} 的支链异构体，命名为异丁烷。

理解丁烷和异丁烷结构之间的差别的最好方法是球-棍模型。

丁烷和异丁烷称为同分异构体的原因是它们之间确实有不同的结构，丁烷包括两个 CH_3 基团和两个 CH_2 基团，而异丁烷包含三个 CH_3 基团和一个 CH 基团。

4. 结构

因为化合物通常是在自然界中首次发现，而不是在实验室中人为制造的。人们可能知道化合物存在，甚至知道它的用处，但无法确切知道原子是由什么组成的，以及它是如何排布的。有几种方法来确定的一种未知化合物的结构：

质谱分析；

X 射线衍射；

核磁共振光谱法；

红外光谱法。

Further Reading
Introduction of Organic Compound

An organic compound is any member of a large class of gaseous, liquid, or solid chemical compounds whose molecules contain carbon. For historical reasons discussed below, a few types of carbon-containing compounds, such as carbides, carbonates, simple oxides of carbon (such as CO and CO_2), and cyanides are considered inorganic. The distinction between organic and inorganic carbon compounds, while "useful in organizing the vast subject of chemistry... is somewhat arbitrary". Organic chemistry is the science concerned with all aspects of organic compounds. Organic synthesis is the methodology of their preparation.

There is no single "official" definition of an organic compound. Some textbooks define an organic compound as one that contains one or more C—H bonds. Others include C—C bonds in the definition. Others state that if a molecule contains carbon——it is organic.

Even the broader definition of "carbon-containing molecules" requires the exclusion of carbon-containing alloys (including steel), a relatively small number of carbon-containing compounds, such as metal carbonates and carbonyls, simple oxides of carbon and cyanides, as well as the allotropes of carbon and simple carbon halides and sulfides, which are usually considered inorganic.

The "C-H" definition excludes compounds that are historically and practically considered organic. Neither urea nor oxalic acid is organic by this definition, yet they were two key compounds in the vitalism debate. The IUPAC Blue Book on organic nomenclature specifically mentions urea and oxalic acid. Other compounds lacking C—H bonds that are also traditionally considered organic include benzenehexol, mesoxalic acid, and carbon **tetrachloride**. Mellitic acid, which contains no C—H bonds, is considered a possible organic substance in Martian soil. C—C bonds are found in most organic compounds, except some small molecules like methane and methanol, which have only one carbon atom in their structure.

The "C—H bond-only" rule also leads to somewhat arbitrary divisions in sets of carbon-fluorine compounds, as, for example, Teflon is considered by this rule to be "inorganic", whereas Tefzel is considered to be organic. Likewise, many Halons are considered inorganic, whereas the rest are considered organic. For these and other reasons, most sources believe that C—H compounds are only a subset of "organic" compounds.

In **summary,** most carbon-containing compounds are organic, and almost all organic compounds contain at least a C—H bond or a C—C bond. A compound does not need to contain C—H bonds to be considered organic (e.g., urea), but many organic compounds do.

New Words

tetrachloride [tetrəˈklɔːraɪd] *n.* 四氯化物

summary [ˈsʌməri] *adj.* 简易的；扼要的 *n.* 概要，摘要，总结

参考文献

[1] Borisover M. The differential Gibbs free energy of sorption of an ionizable organic compound: eliminating the contribution of solute-bulk solvent interactions. Adsorption-journal of the International Adsorption Society, 2016:1-9.

[2] Ehsani A, Shiri H M, et al. Theoretical, common electrochemical and electrochemical noise investigation of inhibitory effect of new organic compound nanoparticles in the corrosion of stainless steel in acidic solution. Transactions of the Indian Institute of Metals, 2016:1-9.

[3] Teymouri M, Karkhane M, et al. Designing a Response Surface Model for

Removing Phosphate and Organic Compound from Wastewater by Pseudomonas Strain MT1. Proceedings of the National Academy of Sciences, India-Section B: Biological Sciences, 2016:1-10.

[4] Chen L, Ludviksson A. Neutral beam etching of cu-containing layers in an organic compound gas environment. WO/2016/003591. 2016.

Unit 2　Biochemistry　生物化学

Chapter 1　Chemical Elements of Life　生命的化学元素

1. The origin of life

　　People have always pondered the riddle of their existence. Indeed, all known cultures, past and present, primitive and sophisticated, have some sort of a creation myth that rationalizes how life arose. Only in the modern era, however, has it been possible to consider the origin of life in terms of a scientific framework, that is, in a manner subject to **experimental** verification.

　　Biochemistry aims to explain **biological** form and function in chemical terms. As we noted earlier, one of the most fruitful approaches to understanding biological phenomena has been to purify an individual chemical component, such as a protein, from a living **organism** and to characterize its structural and chemical characteristics. By the late eighteenth century, chemists had concluded that the composition of living matter is strikingly different from that of the **inanimate** world. Antoine Lavoisier (1743—1794) noted the relative chemical simplicity of the "mineral world" and contrasted it with the complexity of the "plant and animal worlds"; the latter, he knew, were composed of compounds rich in the elements carbon, oxygen, nitrogen, and **phosphorus.**

　　During the first half of the twentieth century, parallel biochemical investigations of **glucose** breakdown in **yeast** and in animal muscle cells

New Words and Expressions

experimental [ɪkˌsperɪˈmentəl]
adj. 实验的；根据实验的；试验性的

biochemistry [ˌbaɪəʊˈkemɪstrɪ]
n. 生物化学

biological [ˌbaɪəˈlɒdʒɪkl]
adj. 生物的；生物学的

organism [ˈɔːgənɪzəm]
n. 有机体；生物体；微生物

inanimate [ɪnˈænɪmət]
adj. 无生命的；无生气的

phosphorus [ˈfɒsfərəs]
n. 磷

glucose [ˈgluːkəʊs]
n. 葡萄糖；葡糖

yeast [jiːst]
n. 酵母

revealed remarkable chemical similarities in these two apparently very different cell types; the breakdown of glucose in yeast and muscle cells involved the same ten chemical intermediates. Subsequent studies of many other biochemical processes in many different organisms have confirmed the generality of this observation, neatly summarized by Jacques Monod: "What is true of *E. coli* is true of the elephant." The current understanding that all organisms share a common evolutionary origin is based in part on this observed universality of chemical intermediates and **transformations**.

Only about 30 of the more than 90 naturally occurring chemical elements are essential to organisms. Most of the elements in living matter have relatively low atomic numbers; only five have atomic numbers above that of selenium, 34 (Figure. 2-1). The four most abundant elements in living organisms, in terms of percentage of total number of atoms, are hydrogen, oxygen, nitrogen, and carbon, which together make up more than 99% of the mass of most cells. They are the lightest elements capable of forming one, two, three, and four bonds, respectively; in general, the lightest elements form the strongest bonds. The trace elements (Figure. 2-1) represent a miniscule fraction of the weight of the human body, but all are essential to life, usually because they are essential to the function of specific proteins, including **enzymes**. The oxygen-transporting capacity of the **hemoglobin** molecule, for example, is absolutely dependent on four iron ions that make up only 0.3% of its mass.

transformation [trænsfəˈmeɪʃən]
n. 转化；转换

enzyme [ˈenzaɪm]
n. 酶
hemoglobin [ˌhiːməʊˈgləʊbɪn]
n. 血红蛋白

1 H																	2 He
3 Li	4 Be		☐ Bulk elements									5 B	6 C	7 N	8 O	9 F	10 Ne
11 Na	12 Mg		☐ Trace elements									13 Al	14 Si	15 P	16 S	17 Cl	18 Ar
19 K	20 Ca	21 Sc	22 Ti	23 V	24 Cr	25 Mn	26 Fe	27 Co	28 Ni	29 Cu	30 Zn	31 Ga	32 Ge	33 As	34 Se	35 Br	36 Kr
37 Rb	38 Sr	39 Y	40 Zr	41 Nb	42 Mo	43 Tc	44 Ru	45 Rh	46 Pd	47 Ag	48 Cd	49 In	50 Sn	51 Sb	52 Te	53 I	54 Xe
55 Cs	56 Ba		72 Hf	73 Ta	74 W	75 Re	76 Os	77 Ir	78 Pt	79 Au	80 Hg	81 Tl	82 Pb	83 Bi	84 Po	85 At	86 Rn
87 Fr	88 Ra		Lanthanides														
			Actinides														

Figure 2-1   Elements essential to animal life and health

2. The chemical elements of life

Living matter, as Table 2-1 indicates, consists of a relatively small number of elements. C, H, O, N, P, and S, all of which readily form covalent bonds, comprise 92% of the dry weight of living things (most organisms are 70% water). The balance consists of elements that are mainly present as ions and for the most part occur only in trace quantities (they usually carry out their functions at the active sites of enzymes). Note, however, that there is no known biological requirement for 64 of the 90 naturally occurring elements Conversely, with the exceptions of oxygen and calcium, the biologically most abundant elements are but minor constituents of **Earth's crust** (the most abundant components of which are O, 47%; Si, 28%; Al, 7.9%; Fe, 4.5%; and Ca, 3.5%).

earth's crust ['ɜːs krʌt] *n.* 地壳；地核

Table 2-1   Elemental Composition of the Human Body

Element	Dry Weight /%	Elements Present in Trace Amounts
C	61.7	B
N	11.0	F
O	9.3	Si
H	5.7	V
		(continued)

Element	Dry Weight /%	Elements Present in Trace Amounts
Ca	5.0	Cr
P	3.3	Mn
K	1.3	Fe
S	1.0	Co
Cl	0.7	Ni
Na	0.7	Cu
Mg	0.3	Zn
		Se
		Mo
		Sn
		SN
		I

参考译文

1. 生命的起源

人们一直以来都在思考人类存在之谜。事实上，所有已知的文化，无论是过去还是现在，原始的还是复杂的，多少都有些使生命如何产生成为合理化而创造的神话。然而，在现代只有按照科学的方法，并经过实验的验证，才有可能思考出生命的起源问题。

生物化学旨在运用化学术语阐释生命的形式和功能。正如我们前面提到的，其中最富有成效的方法来理解生物现象的途径是纯化生物个体的化学成分，如从生物体中提取蛋白质以描述其结构和化学特征。到18世纪末，化学家们已经得出结论，有生命物质的成分明显不同于无生命的世界。安东尼·拉瓦锡（1743—1794）提出了化学上相对简单的"矿物世界"并将它与"植物和动物世界"的复杂性进行对比；他发现，后者是由富含碳、氧、氮和磷元素组成的化合物构成的。

20世纪上半叶，在两种截然不同的酵母和动物细胞中并行开展葡萄糖分解的研究显示出显著的化学相似性。在酵母和肌肉细胞中，葡萄糖的分解都包括相似的十种化学中间产物。在不同的生物体中后续许多其他生化过程的研究已经证实这一观察的普遍性，雅克·莫诺果断地总结出："大肠杆菌的真正本质是什么？大象的真正本质是什么？"目前的理解是，所有的生物共享一个共同的进化起源，这一观点是部分基于已观察到的化学中间体的普遍性和化学中间体之间的转换。

在90多种天然存在的化学元素中，只有大约30种化学元素对生物体来说是必不可少的。大多数生活物质中的元素，其原子序数都较低，只有五种元素的原子序数超过了硒（34）元素（图2-1）。生物体的所有原子中，含量所占比例较多的四种元素是：氢、氧、氮、碳，这四种元素的含量总和在大多数细胞超过了其99%的质量。它们分别是能够形成化学键的元素里，原子序数最小的一个；一般来说，原子序数最小的元素能够形成最牢固的化学键。微量元素（图2-1）只占人体重量中很少的一部分，但都是生命所必不可少的，因为通常它们对特定的蛋白质具有（包括酶）至关重要的作用。例如，血红蛋白分子运输氧的能力完全依赖于它的组成成分中的四个铁离子，而其只占0.3%的质量。

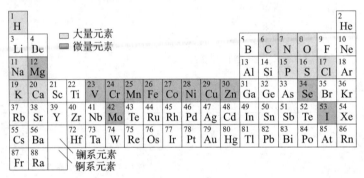

图2-1 动物生命和健康的重要元素

2. 生命的化学组成

生活物质，如表2-1所示为较少的几种元素，C、H、O、N、P和S，这些元素易于形成共价键，占生物体干重的92%（大多数生物干重的70%是水）。元素组成的平衡主要表现为离子的平衡，大部分离子的含量为痕量（他们通常在酶的活性中心部位发挥功能）。然而值得注意的是，在90种天然元素中的64种元素，没有任何已知的生物学含量要求，相反，除了氧元素和钙元素外，生物体中的绝大多数大量元素，在地壳的元素组成中含量较少（最丰富的元素是O，47%；Si，28%；Al，7.9%；Fe，4.5%；和Ca，3.5%）。

表2-1　　　　　　　　　　人体的元素组成

大量元素	干重 /%	微量元素
C	61.7	B
N	11.0	F
O	9.3	Si
H	5.7	V

续表

大量元素	干重 /%	微量元素
Ca	5.0	Cr
P	3.3	Mn
K	1.3	Fe
S	1.0	Co
Cl	0.7	Ni
Na	0.7	Cu
Mg	0.3	Zn
		Se
		Mo
		Sn
		SN
		I

Further Reading

Solid Material in Cells

Most of the solid material of cells consists of carbon-containing compounds. The study of such compounds falls into the domain of organic chemistry. There is considerable overlap between the disciplines of organic chemistry and biochemistry, and a course in organic chemistry is helpful in understanding biochemistry. Organic chemists are more interested in reactions that take place in the laboratory, whereas biochemists would like to understand how reactions occur in living cells.

Biochemical reactions involve specific chemical bonds or parts of molecules called functional groups. Note that all of these linkages consist of several different atoms and individual bonds between atoms. We will encounter these compounds, functional groups, and linkages throughout this book. Ester and ether linkages are common in fatty acids and lipids. Amide linkages are found in proteins. Phosphate ester and phosphoanhydride linkages occur in **nucleotides**.

An important theme of biochemistry is that the chemical reactions that occur inside cells are the same kinds of reactions that take place in a chemistry laboratory. The most important difference is that almost all reactions that occur in living cells are **catalyzed** by enzymes and thus proceed at very high rates. One of the main goals of this textbook is to explain how enzymes speed up reactions without violating the fundamental reaction mechanisms of organic chemistry. The catalytic efficiency of enzymes can be observed even when the enzymes and reactants are isolated in a test tube. Researchers often find it useful to distinguish between biochemical reactions that take place in an organism (*in vivo*) and those that occur under laboratory conditions (*in vitro*).

New Words

nucleotide ['njuːklɪətaɪd] *n.* 核苷；核苷酸

catalyze ['kætəlaiz] *vt.* 催化；刺激，促进

参考文献

[1] Ohno H, Matsubae K, et al. Unintentional Flow of Alloying Elements in Steel during Recycling of End-of-Life Vehicles. Journal of Industrial Ecology, 2014, 18(2):242-253.

[2] Pretty I A. The life course, care pathways and elements of vulnerability. A picture of health needs in a vulnerable population. Gerodontology, 2014, 31(Supplement s1):1-8.

[3] Yano J, Muroi T, et al. Rare earth element recovery potentials from end-of-life hybrid electric vehicle components in 2010-2030. Journal of Material Cycles and Waste Management, 2016:1-10.

[4] Virdun C, Luckett T, et al. Dying in the hospital setting: A systematic review of quantitative studies identifying the elements of end-of-life care that patients and their families rank as being most important. Palliative Medicine, 2015, 29(9):774-796.

Chapter 2 Biomolecules: Carbohydrates, Lipids and Protein 生物分子：糖类、脂类、蛋白质

1. Many important **macromolecules** are **polymers**

Chemical structures are the vocabulary of biochemistry. We present some of these structures to prepare you for the examples you will encounter in the next few chapters. Much of biochemistry deals with very large molecules that we refer to as macromolecules. Biological macromolecules are usually a form of polymer created by joining many smaller organic molecules, or **monomers**, via condensation (removal of the elements of water). Each monomer incorporated into a macromolecular chain is termed a **residue**. In some cases, such as certain **carbohydrates**, a single residue is repeated many times; in other cases, such as proteins and nucleic acids, a variety of residues are connected in a particular order. Each residue of a given polymer is added by repeating the same enzyme-catalyzed reaction. Thus, all of the residues in a macromolecule are aligned in the same direction, and the ends of the macromolecule are chemically distinct.

Macromolecules have properties that are very different from those of their constituent monomers. For example, starch is not **soluble** in water and does not taste sweet, although it is a polymer of the sugar glucose. The organisms were classified as: atoms, molecules, macromolecules, organelles, cells, **tissues**, **organs**, and whole organisms. The following sections briefly describe the principal types of macromolecules and how their sequences of residues or **three-dimensional** shapes grant them unique properties.

New Words and Expressions

macromolecule [mækrəʊ'mɒlɪkjuːl]
n. 高分子；大分子
polymer ['pɒlɪmə]
n. 聚合物

monomer ['mɒnəmə]
n. 单体；单元结构

residue ['rezɪdjuː]
n. 残基；残渣；剩余；滤渣
carbohydrate [kɑːbə'haɪdreɪt]
n. 碳水化合物；糖类

soluble ['sɒljʊbəl]
adj. 可溶的，可溶解的

tissue ['tɪʃuː] *n.* 组织
organ ['ɔːgən]
n. 器官；机构
three-dimensional [θriːdɪmenʃənəl]
adj. 三维的

2. Carbohydrates

Carbohydrates, or **saccharide**, are composed primarily of carbon, oxygen, and hydrogen. This group of compounds includes simple sugars (**monosaccharides**) as well as their polymers (**polysaccharides**). All monosaccharides and all residues of polysaccharides contain several **hydroxyl groups** and are therefore polyalcohols. The most common monosaccharides contain either five or six carbon atoms.

Carbohydrates are **aldehyde** or **ketone** compounds with multiple hydroxyl groups. They make up most of the organic matter on Earth because of their extensive roles in all forms of life. First, carbohydrates serve as energy stores, fuels, and metabolic intermediates. Second, ribose and **deoxyribose sugar** form part of the structural framework of RNA and DNA. Third, polysaccharides are structural elements in the cell walls of **bacteria** and plants. In fact, **cellulose**, the main constituent of plant cell walls, is one of the most abundant organic compounds in the **biosphere**. Fourth, carbohydrates are linked to many proteins and **lipids**, where they play key roles in mediating interactions among cells and interactions between cells and other elements in the cellular environment.

A key related property of carbohydrates in their role as **mediator** of cellular interactions is the tremendous structural diversity possible within this class of molecules. Carbohydrates are built from monosaccharides, small molecules that typically contain from three to nine carbon atoms and vary in size and in the **stereochemical configuration** at one or more carbon centers.

saccharide ['sækəraɪd]
n. 糖；糖类
monosaccharide [mɒnəʊ'sækəraɪd]
n. 单糖
polysaccharide [pɒlɪ'sækəraɪd]
n. 多糖；多聚糖
hydroxyl group [haɪ'drɒksɪl gruːp]
n. 羟基
aldehyde ['ældɪhaɪd]
n. 醛；乙醛
ketone ['kiːtəʊn]
n. 酮
deoxyribose sugar [diːˌɒksɪ'raɪbəʊs 'ʃʊgər]
n. 脱氧核糖
bacteria [bæk'tɪərɪə]
n. 细菌
cellulose ['seljʊləʊz]
n. 纤维素
biosphere ['baɪəʊsfɪə]
n. 生物圈
lipid ['lɪpɪd]
n. 脂质；油脂
mediator ['miːdɪeɪtə]
n. 中介物
stereochemical [sterɪəʊ'kemɪk]
adj. 立体化学的
configuration [kənˌfɪgə'reɪʃən]
n. 配置；结构

These monosaccharides may be linked together to form a large variety of oligosaccharide structures. The unraveling of these oligosaccharide structures, the discovery of their placement at specific sites within proteins, and the determination of their function are tremendous challenges in the field of proteomics.

Glucose is the most abundant six carbon sugar (Figure 2-2a). It is the **monomeric unit** of cellulose, a structural polysaccharide, and of glycogen and starch, which are storage polysaccharides. In these polysaccharides, each glucose residue is joined covalently to the next by a covalent bond between C-1 of one glucose molecule and one of the hydroxyl groups of another. This bond is called a **glycosidic bond**. In cellulose, C-1 of each glucose residue is joined to the C-4 hydroxyl group of the next residue (Figure 2-2b). The hydroxyl groups on adjacent chains of cellulose interact noncovalently, creating strong, insoluble fibers. Cellulose is probably the most abundant biopolymer on Earth because it is a major component of flowering plant stems, including tree trunks.

monomeric unit
[ˌmɔnə'merik juːnit]
n. 单体单元

glycosidic bond
[ˌglaikəu'sidik bɔnd]
n. 糖苷键

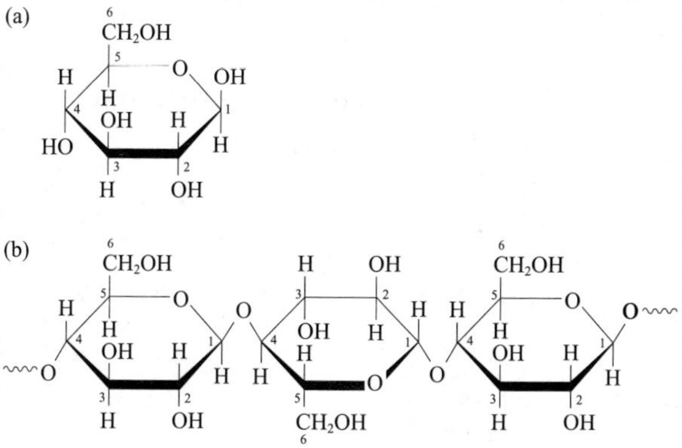

Figure 2-2　Glucose and cellulose

3. Lipids

The term "lipid" refers to a diverse class of molecules that are rich in carbon and hydrogen but contains relatively few oxygen atoms. Most lipids are not soluble in water, but they do dissolve in some organic solvents. The simplest lipids are **fatty acid**. These are long-chain hydrocarbons with a **carboxylate group** at one end. Fatty acids are commonly found as part of larger molecules called **glycerophospholipids**, which contain glycerol 3-phosphate and two **fatty acyl groups**. Glycerophospholipids are major components of biological **membranes**.

Lipids often have a **polar**, **hydrophilic** (water-loving) head that can interact with an aqueous environment, and a nonpolar, **hydrophobic** (water-fearing) tail (Figure 2-3). In an aqueous environment, the hydrophobic tails of such lipids associate, producing a sheet called a **lipid bilayer**. Lipid bilayers form the structural basis of all biological membranes. Membranes separate cells or compartments within cells from their environments by acting as barriers that are impermeable to most water-soluble compounds. Membranes are flexible because lipid bilayers are stabilized by noncovalent forces.

fatty acid ['fætiː æsid]
n. 脂肪酸
carboxylate group [kɑːˈbɒksɪleɪt gruːp]
n. 羧基
glycerophospholipids [gliːsiərəfəʊspˈhɒlɪpɪds]
n. 甘油磷脂
fatty acyl groups ['fætiː 'æsil gruːps]
n. 脂酰基
membrane ['mɛmbren]
n. 膜；薄膜
polar ['pəʊlə]
adj. 极性的；两极的；正好相反的
hydrophilic [haɪdrəʊˈfɪlɪk]
adj. 亲水的
hydrophobic [haɪdrəʊˈfəʊbɪk]
adj. 疏水的
lipid bilayer ['lɪpɪd ˌbaiˈleiə]
n. 磷脂双分子层

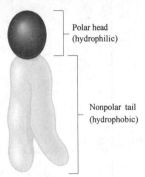

Figure 2-3　Model of a membrane lipid

4. Protein

Twenty common amino acids are incorporated into proteins in all cells. Each amino acid contains an amino group and a carboxylate group, as well as a side chain (R group) that is unique to each amino acid (Figure 2-4a). The amino group of one amino acid and the carboxylate group of another are condensed during protein synthesis to form an amide linkage as shown in Figure 2-4b. The bond between the carbon atom of one amino acid residue and the nitrogen atom of the next residue is called a peptide bond. The end-to-end joining of many amino acids forms a linear **polypeptide** that may contain hundreds of amino acid residues. A functional protein can be a single polypeptide, or it can consist of several different polypeptide chains that are tightly bound to form a more complex structure.

polypeptide [ˌpɒlɪˈpeptaɪd]
n. 多肽，缩多氨酸

(a) $H_3N^+\!-\!\underset{\underset{R}{|}}{\overset{\overset{COO^-}{|}}{C}}\!-\!H$

(b) $H_3N^+\!-\!\underset{\underset{R}{|}}{CH}\!-\!\overset{\overset{O}{\|}}{C}\!-\!\underset{\underset{H}{|}}{N}\!-\!\underset{\underset{R}{|}}{CH}\!-\!COO^-$

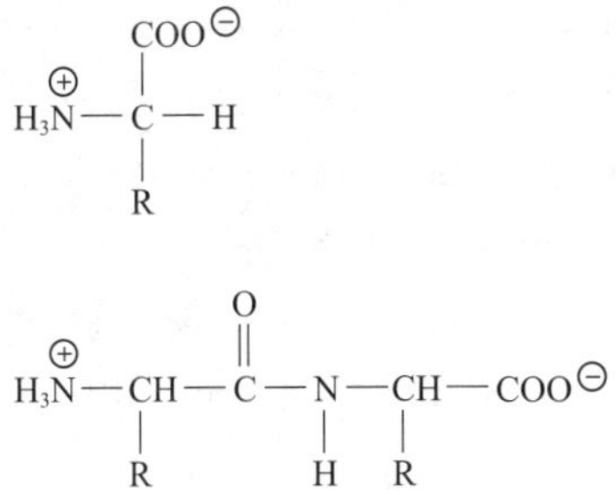

Figure 2-4 Structure of an amino acid and a **dipeptide**

dipeptide [daɪˈpepˌtaɪd]
n. 二肽

Proteins are the most versatile macromolecules in living systems and serve crucial functions in essentially all biological processes. They function as catalysts, they transport and store other molecules such as oxygen, they provide mechanical support and immune protection, they generate movement, they transmit **nerve** impulses, and they control growth and **differentiation**.

nerve [nɜːv]
n. 神经
differentiation [ˌdɪfərenʃɪˈeɪʃn]
n. 分化

Proteins are linear polymers built of monomer units called amino acids. The construction of a vast array of macromolecules from a limited number of monomer building blocks is a recurring theme in biochemistry. Does protein function depend on the linear sequence of amino acids? The function of a protein is directly dependent on its three dimensional structure. Remarkably, proteins spontaneously fold up into three-dimensional structures that are determined by the sequence of amino acids in the protein polymer. Thus, proteins are the embodiment of the transition from the one-dimensional world of sequences to the three-dimensional world of molecules capable of diverse activities.

Proteins contain a wide range of functional groups. These functional groups include **alcohols**, **thiols**, **thioethers**, **carboxylic acids**, **carboxamides**, and a variety of basic groups. When combined in various sequences, this array of functional groups accounts for the broad spectrum of protein function. For instance, the chemical **reactivity** associated with these groups is essential to the function of enzymes, the proteins that catalyze specific chemical reactions in biological systems.

Proteins can interact with one another and with other biological macromolecules to form complex assemblies. The proteins within these assemblies can act synergistically to generate capabilities not afforded by the individual component proteins. These assemblies include macro-molecular machines that carry out the accurate replication of DNA, the transmission of signals within cells, and many other essential processes.

alcohols [ˈælkəˌhɔl]
n. 醇类
thiols [θiˈolz] n. 硫醇
thioethers [θaiəuˈiːθə]
n. 硫醚
carboxylic acids
[ˌkɑːbɔkˈsilik ˈæsids]
n. 羧酸
carboxamide
[kɑːˈbɔksæmaid]
n. 甲酰胺；酰胺；羧胺
reactivity [riækˈtɪvɪti]
n. 反应

参考译文

1. 许多重要的高分子聚合物

化学结构是生物化学的核心词汇。我们先为大家准备几个在接下来几章中即将遇到的结构的例子。生物化学中的大部分内容与大分子有关,我们称之为生物大分子。生物大分子通常是由很多小分子有机物或单体通过缩合(脱去水分子),形成的聚合物形式。每一个加入到大分子链的单体称为残基。在某些情况下,如某些碳水化合物,一个残基被多次重复;在其他情况下,如蛋白质和核酸,各种残基以特定的顺序结合。给定聚合物中每一个残基是由相同的酶促反应重复加入的。因此,大分子中的所有残基在同一个方向上一致,但其末端不同。

大分子的性质不同于组成它的单体。例如,尽管淀粉是由葡萄糖组成的聚合物,但淀粉不溶于水,也没有甜味。生物体的分类如下:原子、分子、高分子、细胞器、细胞、组织、器官和整个生物体。下面部分将简要描述几种大分子的主要类型和他们的残基序列或三维结构如何赋予它们特殊的性质。

2. 糖类

碳水化合物或者称糖类,主要是由碳、氧和氢组成的。这组化合物包括简单的糖(单糖)以及它们的聚合物(多糖)。所有单糖和多糖的残基中含有多个羟基,因此属于多元醇。最常见的单糖含有五个或六个碳原子。

碳水化合物为多羟基的醛或酮的化合物。由于它们广泛地存在于所有形式的生命中,所以它们组成了地球上大部分的有机物质。首先,碳水化合物能够储存能量,作为燃料和代谢的中间体。第二,核糖和脱氧核糖形成了 RNA 和 DNA 的结构骨架部分。第三,多糖是细菌和植物细胞壁的结构原件。事实上,植物细胞壁的主要成分纤维素是生物圈最丰富的有机化合物之一。第四,碳水化合物和许多蛋白质、脂类相连,能够起到调节细胞和细胞之间,以及细胞与细胞环境中其他原件相互作用的作用。

与碳水化合物作为细胞间相互作用的介质相关的一个关键性质,是这类分子极大的结构多样性。碳水化合物有单糖构成,典型的小分包含三到九个碳原子,大小和一个或多个碳中心的立体化学结构各异。这些单糖可以连接在一起形成一个各种低聚糖的结构。解开这些低聚糖的结构,探索它们在蛋白质特异位点的位置,和确定它们的功能都是蛋白质组学领域的巨大挑战。

葡萄糖是最丰富的六元糖(图 2-2a)。它是纤维素的结构单体,也是存储性多糖——糖原和淀粉的结构单体。在这些多糖中,每一个葡萄糖残基以 C-1 和另一个葡萄糖的羟基,通过共价键相连。这个键称为糖苷键。在纤维素中,每一个葡萄糖残基和相邻残基的 C-4 上的羟基相连(图 2-2b)。在相邻羟基纤维素链共价相互作用下,形成难溶性的纤维。纤维素是地球上最丰富的生物聚合物,因为它是开花植物茎和树干的主要组成成分。

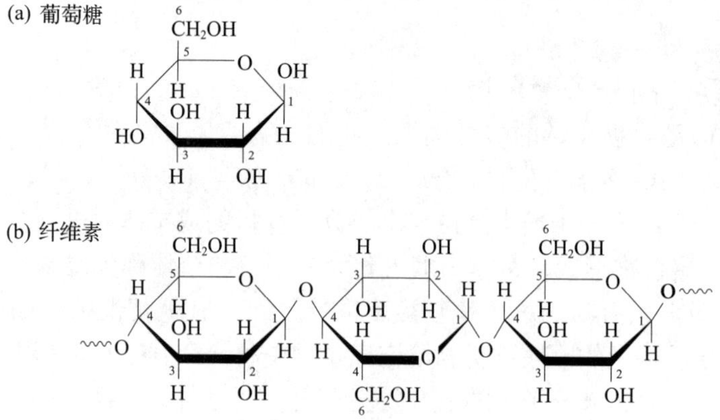

(a) 葡萄糖

(b) 纤维素

图 2-2 葡萄糖和纤维素

3. 脂类

"脂类"一词指的是一种富含碳和氢，但含有氧原子相对较少的不同分子。大多数脂肪不溶于水，但溶于某些有机溶剂。最简单的脂类是脂肪酸。这些长链碳氢化合物的一端均为羧基。脂肪酸中普遍存在的较大的分子称为甘油磷脂，含有 3-磷酸甘油和脂酰 2 个脂酰基。甘油磷脂是生物膜的主要成分。

脂质通常有一个极性亲水（水溶性的）头部，可以与水环境进行相互作用，和一个非极性（疏水性的）尾部（图 2-3）。在水环境中，脂类的疏水尾部产生的片层称为脂质双分子层。脂质双分子层是所有生物膜的结构基础。膜将细胞分离或者在细胞成分与外界环境中形成屏障，以阻止水溶性化合物的渗透。双分子层受共价力作用而稳定。

4. 蛋白质

细胞中的蛋白质含有常见的 20 种氨基酸。每个氨基酸含有一个氨基和一个羧基，以及对于每一个氨基酸特异的侧链（R 基）（图 2-4a）。一个氨基酸的氨基和另一个氨基酸的羧基在蛋白质的合成中，形成酰胺键如图 2-4b 所示。位于一个氨基酸侧链的碳原子和下一个残基的氮原子之间的化学键称为肽键。许多氨基酸首尾相连接形成一个线性多肽可能包含数百个氨基酸残基。有功能的蛋白质可以单个多肽，也可以由几个不同的多肽链紧密地结合成一个更复杂的结构形式。

蛋白质是生命系统中最重要的多功能生物大分子，在所有的生物过程中起到至关重要的作用。它们的功能是作为催化剂、运输和储存分子氧等其他分子、提供机械支撑和免疫保护、产生运动、传递神经冲动，并控制生长和分化。

蛋白质是由称为氨基酸的单体组成的线性聚合物。从有限数量的单体构建出大量的生物大分子是生物化学中常见的构成方式。蛋白质的功能是取决于氨基

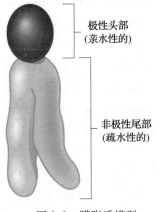

图 2-3　膜脂质模型　　　　图 2-4　氨基酸和二肽的结构

酸的线性序列吗？蛋白质的功能直接由其三维结构决定。值得注意的是，蛋白质自发地折叠成三维结构，这些结构是由氨基酸在蛋白质聚合物的序列决定。因此，蛋白质是将一维序列世界转化为具有活动多样性的三维分子的能力的一种体现。

蛋白质中包含大量功能基团。这些基团包括：醇类、硫醇、硫醚、羧酸、酰胺和很多基本基团。当结合在各种各样不同的序列中，这些功能基团的排列能够产生很多蛋白质的功能。例如，这些与化学反应活性相关的基团是酶的功能所必需的，这些蛋白质在生物系统中催化特异的化学反应。

蛋白质和蛋白质之间，以及和其他生物大分子之间都可发生相互作用，形成复杂的组装。这些组装中的蛋白质可以不受单独的蛋白质成分的影响，而协同作用产生新的能力。这些组装包括能够进行 DNA 精确复制的微分子机器，细胞间信号转导和许多其他重要过程。

Further Reading
Biomolecules and Human Biology

Our understanding of biochemistry has had and will continue to have extensive effects on many aspects of human endeavor. *First, biochemistry is an intrinsically beautiful and fascinating body of knowledge.* We now know the essence and many of the details of the most fundamental processes in biochemistry, such as how a single molecule of DNA replicates to generate two identical copies of itself and how the sequence of bases in a DNA molecule determines the sequence of amino acids in an encoded protein. Our ability to describe these processes in detailed, mechanistic terms places a firm chemical foundation under other biological sciences. Moreover, the realization that we can understand essential life processes, such as the transmission of hereditary information, as chemical structures and their reactions has significant philosophical implications. What does it mean, biochemically, to be human? What are the biochemical differences between a human being, a chimpanzee, a mouse, and a fruit fly? Are we more similar than we are different?

Second, biochemistry is greatly influencing medicine and other fields. The molecular lesions causing **sickle-cell anemia, cystic fibrosis, hemophilia**, and many other genetic diseases have been elucidated at the biochemical level. Some of the molecular events that contribute to cancer development have been identified. An understanding of the underlying defects opens the door to the discovery of effective therapies. Biochemistry makes possible the rational design of new drugs, including specific inhibitors of enzymes required for the replication of viruses such as human immunodeficiency virus (HIV). Genetically engineered bacteria or other organisms can be used as "factories" to produce valuable proteins such as insulin and stimulators of blood cell development. Biochemistry is also contributing richly to clinical diagnostics. For example, elevated levels of telltale enzymes in the blood reveal whether a patient has recently had a myocardial infarction (heart attack). DNA probes are coming into play in the precise diagnosis of inherited disorders, infectious diseases, and cancers. Agriculture, too, is benefiting from advances in biochemistry with the development of more effective, environmentally safer herbicides and pesticides and the creation of genetically engineered plants that are, for example, more resistant to insects. All of these endeavors are being accelerated by the advances in genomic sequencing.

Third, advances in biochemistry are enabling researchers to tackle some of the most exciting questions in biology and medicine. How does a fertilized egg give rise to cells as different as those in muscle, brain, and liver? How do the senses work? What are the molecular bases for mental disorders such as Alzheimer disease and schizophrenia? How does the immune system distinguish between self and oneself? What are the molecular mechanisms of short-term and long-term memory? The answers to such questions, which once seemed remote, have been partly uncovered and are likely to be more thoroughly revealed in the near future.

New Words

sickle-cell anemia ['sɪkls'el ə'niːmiːə] *n.* 镰状细胞性贫血

cystic fibrosis ['sɪstɪk ˌfaɪ'brəʊsɪs] *n.* 囊胞性纤维症

hemophilia [hiːmə'fɪlɪə] *n.* 血友病

参考文献

[1] Sudha P N, Gomathi T, et al. Marine carbohydrates of wastewater treatment. Advances in Food & Nutrition Research, 2014, 73(5):103–113.

[2] Helfrich W. Elastic properties of lipid bilayers: theory and possible experiments.

Zeitschrift Fur Naturforschung Teil Biochemie Biophysik Biologie Virologie, 2014, 28(28):693-703.

[3] Bradford M M. A rapid and sensitive method for the quantitation of microgram quantities of protein utilizing the principle of protein-dye binding. Analytical Biochemistry, 2015, 72(s 1-2):248-254.

[4] Mumper R J, Rolland A. Chitosan and chitosan oligomers for nucleic acid delivery: Original research article: Chitosan and depolymerized chitosan oligomers as condensing carriers for in vivo plasmid delivery, 1998. Journal of Controlled Release, 2014, 190:46-48.

Chapter 3 Glycolysis 糖酵解

The first **metabolic** pathway that we encounter is **glycolysis**, an ancient pathway employed by a host of organisms. Glycolysis is the sequence of reactions that metabolizes one molecule of glucose to two molecules of **pyruvate** with the concomitant net production of two molecules of ATP. Pyruvate can be further processed **anaerobically** (fermented) to **lactate** (lactic acid **fermentation**) or **ethanol** (alcoholic fermentation). Under aerobic conditions, pyruvate can be completely oxidized to CO_2, generating much more ATP.

Our understanding of glucose metabolism, especially glycolysis, has a rich history. Indeed, the development of biochemistry and the delineation of glycolysis went hand in hand. A key discovery was made by Hans Buchner and Eduard Buchner in 1897, quite by accident. The Buchners were interested in manufacturing cell-free extracts of yeast for possible therapeutic use. These extracts had to be preserved without the use of **antiseptics** such as **phenol**, and so they decided to try **sucrose**, a commonly used preservative in kitchen chemistry. They obtained a startling result: sucrose was rapidly fermented into alcohol by the yeast juice. The significance of this finding was immense. The Buchners demonstrated for the first time that fermentation could take place outside living cells. The accepted view of their day, asserted by **Louis Pasteur** in 1860, was that fermentation is inextricably tied to living cells. The chance discovery of the Buchners refuted this vitalistic dogma and opened the door to modern biochemistry.

New Words and Expressions

metabolic [metə'bɒlɪk]
adj. 变化的；新陈代谢的
glycolysis [glaɪ'kɒlɪsɪs]
n. 糖酵解
pyruvate [paɪ'ruːveɪt]
n. 丙酮酸
anaerobically [ˌæneərəʊbɪkəlɪ]
adv. 厌氧地
lactate [læk'teɪt]
n. 乳酸
fermentation [fɜːmen'teɪʃən]
n. 发酵
ethanol ['eθənɒl]
n. 乙醇，酒精

antiseptic [æntɪ'septɪk]
n. 防腐剂，抗菌剂
phenol ['fiːnɒl]
n. 苯酚
sucrose ['sjʊkros]
n. 蔗糖

Louis Pasteur
路易•巴斯德（法国化学家及微生物学家）

1. Glycolysis pathway in many organisms

We now start our consideration of the **glycolytic** pathway. This pathway is common to virtually all cells, both **prokaryotic** and **eukaryotic**. In eukaryotic cells, glycolysis takes place in the cytosol. This pathway can be thought of as comprising three stages (Figure 2-5). Stage 1, which is the conversion of glucose into **fructose** 1,6-bisphosphate, consists of three steps: a **phosphorylation**, an **isomerization**, and a second phosphorylation reaction. The strategy of these initial steps in glycolysis is to trap the glucose in the cell and form a compound that can be readily cleaved into phosphorylated three-carbon units. Stage 2 is the cleavage of the fructose 1,6-bisphosphate into two three-carbon fragments. These resulting three-carbon units are readily interconvertible. In stage 3, ATP is harvested when the three-carbon fragments are oxidized to pyruvate.

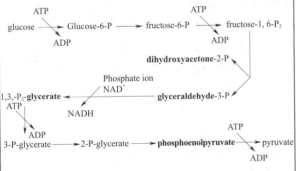

Figure 2-5 Glycolytic pathway

There are ten important enzymes in this reaction, they are **hexokinase, phosphoglucose isomerase**, PFK-1 which is short for phosphofructokinase-1, **aldolase**, triose phosphate isomerase, glyceraldehyde 3-phosphate **dehydrogenase**, phosphoglycerate kinase, **phosphoglycerate mutase**, **enolase**, pyruvate kinase.

The glycolysis starts with glucose under the condition lacking or without O_2, glucose would be **catalyzed** by hexokinase, and glucose would be transformed in glucose-6-phosphate and this process is drived by ATP in a same amount with glucose. Then glucose-6-phosphate would be catalyzed by phosphoglucose isomerase without consuming any ATP, and was transformed into fructose-6-phosphate. PFK-1, which is short for phosphofructokinase-1,is the main enzyme in the next step of transforming fructose-6-phosphate into fructose-1,6-bisphosphate, it will comsume ATP same with the first step. Then fructose-1,6-bisphosphate would be catalyzed by aldolase, glyceraldehydes-3-phosphate and dihydroxyacetone-phosphate in same amount would be produced.

Only glyceraldehydes-3-phosphate can take part in the next steps of glycolysis in animals and human's cell, so dihydroxyacetone-phosphate would be transformed into glyceraldehydes-3-phosphate catalyzed by triose phosphate isomerase. So 1 mol glucose would be transformed into 2 mol glyceraldehydes-3-phosphate. This is the first stage of glycosis which would consume 2 mol ATP without producing any energy. But the next stage will be a producing stage. In this stage, glyceraldehydes-3-phosphate would be catalyzed by glyceraldehyde 3-phosphate dehydrogenase and it would be transformed into 1,3-biophosphoglycerate, and then phosphoglycerate kinase would catalyze it into 3-phosphoglycerate and the same amount of ADP would be transformed into ATP. 3-phosphoglycerate would firstly transformed

catalyze ['kætəlaɪz]
vt. 催化

into 2-phosphoglycerate by phosphoglycerate mutase without consuming or producing any ATP. 2-phosphoglycerate then transformed into phosphoenolpyruvate catalyze by enolase.

Then in the last step, phosphoenolpyruvate would be catalyzed by pyruvate kinase and was transformed into pyruvate with same amount ATP was produced.

The whole reaction can be written as:

Glucose+2NAD$^+$+ADP+2Pi⟶2Pyruvate+2NADH+2ATP+2H$^+$+2H$_2$O

In a word, 1 Glucose is converted to 2 pyruvate and yields 2 ATP.

After this ten steps, pyruvate would get into the **citric acid cycle**, and was transformed into lactate and ethanol.

2. The citric acid cycle

Glucose can be metabolized to pyruvate anaerobically to synthesize ATP through the glycolytic pathway. Glycolysis, however, harvests but a fraction of the ATP available from glucose. We now begin an exploration of the aerobic processing of glucose, which is the source of most of the ATP generated in metabolism. The aerobic processing of glucose starts with the complete **oxidation** of glucose derivatives to carbon dioxide. This oxidation takes place in the citric acid cycle, a series of reactions also known as the **tricarboxylic acid** (TCA) cycle or the **Krebs cycle**. The citric acid cycle is the final common pathway for the oxidation of fuel molecules amino acids, fatty acids, and carbohydrates. Most fuel molecules enter the cycle as **acetyl coenzyme A**.

The citric acid cycle is the central metabolic hub of the cell. It is the gateway to the aerobic

citric acid cycle
['sɪtrɪk, 'æsɪd, 'saɪkəl]
n. 柠檬酸循环

oxidation [ɒksɪ'deɪʃən]
n. 氧化
tricarboxylic acid
[traɪˌkɑːbɒk'sɪlɪk 'æsɪd]
n. 三羧酸
Krebs cycle [krebz 'saɪkl]
n. 克雷布斯循环
acetyl coenzyme A
['æsɪtaɪl 'kəʊˌenzaɪm A]
n. 乙酰辅酶 A

metabolism of any molecule that can be transformed into an **acetyl group** or **dicarboxylic acid**. The cycle is also an important source of precursors, not only for the storage forms of fuels, but also for the building blocks of many other molecules such as amino acids, nucleotide bases, **cholesterol**, and **porphyrin** (the organic component of heme).

What is the function of the citric acid cycle in transforming fuel molecules into ATP? Recall that fuel molecules are carbon compounds that are capable of being oxidized of losing **electrons**. The citric acid cycle includes a series of oxidation-reduction reactions that result in the oxidation of an acetyl group to two molecules of carbon dioxide.

The overall pattern of the citric acid cycle is shown in Figure 2-6. A four carbon compound (**oxaloacetate**) condenses with a two-carbon acetyl unit to yield a six-carbon tricarboxylic acid (citrate). Oxygen is required for the citric acid cycle indirectly inasmuch as it is the electron acceptor at the end of the electron-transport chain, necessary to regenerate NAD$^+$ and FAD.

acetyl group ['æsitil gruːp]
n. 乙酰基
dicarboxylic acid [daiˌkɑːbɔk'silik 'æsid]
n. 二羧酸
cholesterol [kə'lestərɒl]
n. 胆固醇
porphyrin ['pɔːfɪrɪn]
n. 卟啉

electron [ɪ'lektrɒn]
n. 电子

oxaloacetate [ɑksəlo'æsɪˌtet]
n. 草酰乙酸

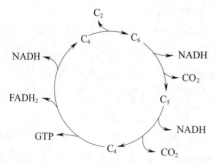

Figure 2-6　Overview of the Citric Acid Cycle

The citric acid cycle, in conjunction with **oxidative phosphorylation**, provides the vast

oxidative phosphorylation ['ɒksɪdeɪtɪv ˌfɔˌsfɔri'leɪʃən]
n. 氧化磷酸化

majority of energy used by aerobic cells in human beings, greater than 95%. It is highly efficient because a limited number of molecules can generate large amounts of NADH and FADH$_2$. Note in Figure 2-6 that the four-carbon molecule, oxaloacetate, that initiates the first step in the citric acid cycle is regenerated at the end of one passage through the cycle. The oxaloacetate acts catalytically: it participates in the oxidation of the acetyl group but is itself regenerated. Thus, one molecule of oxaloacetate is capable of participating in the oxidation of many acetyl molecules.

参考译文

糖酵解是我们遇到的第一个代谢途径，一个古老的存在于许多生物中的代谢通路。糖酵解是将一分子的葡萄糖分解成两分子丙酮酸的一系列反应，并伴随产生两个分子ATP。这个过程是厌氧的（即不需要O_2）。丙酮酸可以进一步厌氧代谢（发酵）生成乳酸（乳酸发酵）和乙醇（酒精发酵）。在有氧条件下，丙酮酸可以被完全氧化成二氧化碳，生成更多的ATP。

我们对葡萄糖代谢的理解，尤其是糖酵解，具有丰富的历史。事实上，生物化学是和糖酵解的发现一起发展起来的。关键的一个是1897年由汉斯·毕希纳和爱德华·毕希纳偶然发现的。毕希纳对提取游离酵母而用于可能的治疗用途很感兴趣。这些提取必须保持酵母的活性，而不使用防腐剂如苯酚，所以他们决定尝试蔗糖——厨房化学常用的防腐剂。他们获得了一个惊人的结果：酵母的汁能让蔗糖快速发酵成酒精。这一发现的意义是巨大的。毕希纳首次证明发酵可以发生在活细胞以外。那一时期被广泛接受的观点是，1860年路易·巴斯德提出的发酵与活细胞密不可分。毕希纳的发现驳斥了这种活力论的观点，为现代生物化学打开了大门。

1. 糖酵解是存在于许多生物中的一个能量转化途径

我们现在开始探讨糖酵解途径。这个是几乎所有细胞、原核和真核生物常见的途径。在真核细胞中，糖酵解过程发生在细胞质。这个途径可以划分成三个阶段（图2-5）。第一阶段，即葡萄糖转化为果糖1,6-二磷酸，由三个步骤组成：磷酸化，异构化，第二个磷酸化反应。这些初始步骤的策略在于将葡萄糖转化成能够分解成含三个碳的化合物。第二阶段：果糖1,6-二磷酸裂解成两个含三个碳的化合物。这些生成的含三个碳的化合物是可以相互转换的。第三阶段：含三个碳的化合物氧化成丙酮酸，同时生产 ATP。

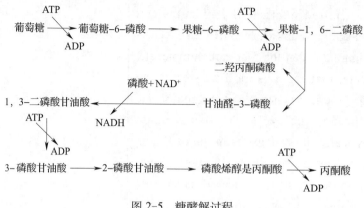

图 2-5 糖酵解过程

有十个重要的酶促反应，它们分别是：己糖激酶、磷酸葡萄糖异构酶（简称为 PFK-1）、醛缩酶、磷酸丙糖异构酶、甘油醛 3－磷酸脱氢酶、磷酸甘油酸酯激酶、磷酸甘油酸酯变位酶、烯醇酶、丙酮酸激酶。

糖酵解的起始是在缺乏或无氧气的条件下，由己糖激酶将葡萄糖转化为葡萄糖-6-磷酸，这一过程是在与葡萄糖等量的 ATP 的驱动下完成。然后葡萄糖-6-磷酸由磷酸葡萄糖异构酶催化，在不消耗任何 ATP 的条件下，转化成果糖-6-磷酸。PFK-1，为磷酸果糖激酶-1 的简称，是下一步将果糖-6-磷酸转化成果糖-1, 6-二磷酸的主要酶，这一过程将消耗与第一步等量的 ATP。然后果糖-1, 6-二磷酸在醛缩酶的催化下生成等量的二羟丙酮磷酸和甘油醛-3-磷酸。

只有甘油醛-3-磷酸可以参与动物和人体中糖酵解的下面的步骤，所以二羟丙酮磷酸将由磷酸丙糖异构酶催化转化成甘油醛-3-磷酸。所以 1mol 葡萄糖会变成 2mol 甘油醛-3-磷酸。因为这是糖酵解的第一阶段，这将消耗 2 摩尔 ATP，但没有产生任何能量。在这个阶段，甘油醛-3-磷酸在甘油醛-3-磷酸脱氢酶催化下，转化成 1, 3-二磷酸甘油酸，然后在磷酸甘油酸激酶催化下生成 3-磷酸甘油酸，同时又有等量的 ADP 转化成 ATP。3-磷酸甘油酸首先要在磷酸甘油酸变位酶催化下转化成 2-磷酸甘油酸，这一过程既没有 ATP 消耗，也没有 ATP 生成。然后，2-磷酸甘油酸在烯醇酶的催化下，转化成磷酸烯醇是丙酮酸。

最后一步，磷酸烯醇是丙酮酸在丙酮酸激酶的催化下转化成丙酮酸，并消耗等量的 ATP。

整个反应过程可以写成：

葡萄糖 +2NAD$^+$+ADP+2Pi⟶2 丙酮酸 +2NADH+2ATP+2H$^+$+2H$_2$O

总而言之，1 分子的葡萄糖转化成 2 分子的丙酮酸，净生成 2 分子的 ATP。经过这十步反应，丙酮酸将进入柠檬酸循环，或转化成乳酸和乙醇。

2. 柠檬酸循环

无氧条件下，葡萄糖通过糖酵解途径代谢生成丙酮酸以产生 ATP。然而，糖酵解只提供出葡萄糖中的一部分 ATP。众所周知，我们开始探索糖的有氧分解，就是因为它是代谢生成 ATP 的主要来源。葡萄糖的有氧分解过程开始于葡萄糖彻底氧化分解成二氧化碳。这一氧化过程发生在柠檬酸循环中，这一系列反应也可称为三羧酸循环（TCA）或 Krebs 循环。柠檬酸循环也是氨基酸、脂肪酸和碳水化合物，这些能量分子的最终共同分解途径。大多数的能量分子都以乙酰 CoA 的形式进入柠檬酸循环。

柠檬酸循环是细胞代谢的中心。它是任何能够转化成酰基和二羧酸的分子，有氧代谢的途径。柠檬酸循环也是前体物质的重要来源，不仅包括能量的贮存形式，而且还构成很多其他分子，如氨基酸、核苷酸碱基、胆固醇和卟啉（亚铁血红素的有机成分）。

柠檬酸循环在将燃料分子转化成 ATP 的过程中的作用是什么？别忘了燃料分子是能够失去电子被氧化的碳化合物。柠檬酸循环包括一系列能将乙酰基氧化成两分子二氧化碳的氧化还原反应。

图 2-6 显示了柠檬酸循环的总体模式。一个四个碳的化合物（草酰乙酸）和一个两个碳的乙酰基单元结合生成一个六个碳的三羧酸（柠檬酸）。氧是三羧酸循环直接必需的，因为它是位于电子传递链末端的电子受体，同时也是 NAD^+ 和 FAD 再生所必需的。

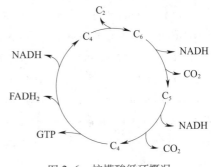

图 2-6 柠檬酸循环概况

柠檬酸循环和氧化磷酸化提供了人类有氧细胞 95% 以上的能量。它是很高效的，因为有限数量的分子能够产生大量的 NADH 和 $FADH_2$。注意一下，图 2-6 中含四个碳的分子，草酰乙酸是柠檬酸循环中起始的第一步通过循环在其末端又重新形成。草酰乙酸的催化行为：它参与乙酰基的氧化分解，而又可自身合成。因此，一分子的草酰乙酸能够参与氧化很多乙酰分子。

Further Reading
Hexokinase and Pyruvate Kinase in Glycolysis

Phosphofructokinase is the most prominent regulatory enzyme in glycolysis, but it is not the only one. Hexokinase, the enzyme catalyzing the first step of glycolysis, is inhibited by its product, glucose 6-phosphate. High concentrations of this molecule signal that the cell no longer requires glucose for energy, for storage in the form of glycogen, or as a source of biosynthetic precursors, and the glucose will be left in the blood.

Why is phosphofructokinase rather than hexokinase the pacemaker of glycolysis? The reason becomes evident on noting that glucose 6-phosphate is not solely a glycolytic intermediate. Glucose 6-phosphate can also be converted into glycogen or it can be oxidized by the pentose phosphate pathway to form NADPH. The first irreversible reaction unique to the glycolytic pathway, the committed step, (Section 10.2), is the phosphorylation of fructose 6-phosphate to fructose 1,6-bisphosphate. Thus, it is highly appropriate for phosphofructokinase to be the primary control site in glycolysis. In general, the enzyme catalyzing the committed step in a metabolic sequence is the most important control element in the pathway.

Pyruvate kinase, the enzyme catalyzing the third irreversible step in glycolysis, controls the outflow from this pathway. This final step yields ATP and pyruvate, a central metabolic intermediate that can be oxidized further or used as a building block. Several **isozymic** forms of pyruvate kinase (a tetramer of 57kd subunits) encoded by different genes are present in mammals: the L type predominates in liver, and the M type in muscle and brain. The L and M forms of pyruvate kinase have many properties in common. Both bind phosphoenolpyruvate cooperatively. Fructose 1,6-bisphosphate, the product of the preceding irreversible step in glycolysis, activates both isozymes to enable them to keep pace with the oncoming high flux of intermediates. ATP allosterically inhibits both the L and the M forms of pyruvate kinase to slow glycolysis when the energy charge is high. Finally, alanine (synthesized in one step from pyruvate) also allosterically inhibits the pyruvate kinases——in this case, to signal that building blocks are abundant.

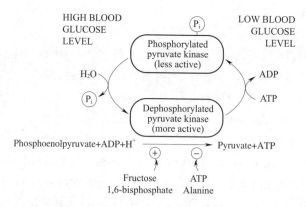

The isozymic forms differ in their susceptibility to covalent modification. The catalytic properties of the L form——but not of the M form——are also controlled by reversible phosphorylation. When the blood-glucose level is low, the **glucagon**-triggered cyclic AMP cascade leads to the phosphorylation of pyruvate kinase, which diminishes its activity. These hormone-triggered phosphorylations, like that of the bifunctional enzyme controlling the levels of fructose 2,6-bisphosphate, prevent the liver from consuming glucose when it is more urgently needed by brain and muscle. We see here a clear-cut example of how isoenzymes contribute to the metabolic diversity of different organs. We will return to the control of glycolysis after considering **gluconeogenesis**.

New Words

phosphofructokinase ['fɒsfəʊˌfrʌktəʊ'kaɪˌneɪs] *n.* 磷酸果糖激酶

isozymic [ˌaɪsəʊ'zaɪmɪk] *adj.* 同工酶的

glucagon ['gluːkəgən] *n.* 胰高血糖素

gluconeogenesis [ˌglʊkoˌnɪə'dʒɛnəsɪs] *n.* 糖异生

参考文献

[1] Roberts D J, Miyamoto S. Hexokinase II integrates energy metabolism and cellular protection: Acting on mitochondria and TORCing to autophagy. Cell Death & Differentiation, 2015, 22(2):248-257.

[2] Dai W, Wang F, et al. By reducing hexokinase 2, resveratrol induces apoptosis in HCC cells addicted to aerobic glycolysis and inhibits tumor growth in mice. Oncotarget, 2015, 6(15):13703-13717.

[3] Guo D, Gu J, et al. Inhibition of pyruvate kinase M2 by reactive oxygen species contributes to the development of pulmonary arterial hypertension. Journal of Molecular & Cellular Cardiology, 2016, 91:179-187.

[4] Ruyters, Grotjohann G, et al. Oligomeric Forms Of Pyruvate Kinase From Chlorella With Different Kinetic Properties. Zeitschrift für Naturforschung C, 2016, 46(5-6):416-422.

Chapter 4   Diabetes and Human Health   糖尿病和人类健康

1. What is **diabetes** mellitus?

   Diabetes mellitus is a kind of metabolic disease that is brought about by either the insufficient production of **insulin** or the inability of the body to respond to the insulin formed within the system. The disease can be classified into two different categories: the type 1and type 2. Diabetes mellitus type 1 is caused by the loss of beta cells found in the islets of Langerhans in the **pancreas**. Another type of diabetes mellitus is the type 2 diabetes. Type 2 diabetes is generally characterized by the body's resistance to insulin. This is primarily attributed to the loss of certain insulin receptors in the tissues that are supposed to mediate the entrance of insulin into the body's cells.

   This particular kind of diabetes is the most common kind that afflicts most of the reported cases of the disease. Type 2 diabetes usually leads to **hyperglycemia** which can be tre Diabetes **Symptoms**. All of these forms of diabetes have different symptoms as well as other different aspects of the disease. They also all have different **complications**. Type 1 diabetes refers to a condition where there is no insulin to metabolize glucose and leads to low levels of glucose in the blood where Type 2 diabetes is a condition where there is abnormally high glucose in the blood.

2. What causes diabetes?

   The causes of diabetes are not clearly known. There is still a mystery in it. Medical scientists are unable to tell, that's why some people suffer from

New Words and Expressions

diabetes [daɪəˈbiːtiːz]
n. 糖尿病；多尿症

insulin [ˈɪnsjʊlɪn]
n. 胰岛素

pancreas [ˈpæŋkrɪəs]
n. 胰腺

hyperglycemia [haɪpəglaɪˈsiːmɪə]
n. 高血糖症

symptom [ˈsɪmptəm]
n. 症状

complication [kɒmplɪˈkeɪʃən]
n. 并发症

diabetes while others do not. There are certain factors which may lead to diabetes. These factors are considered as risk factors and an individual must try to overcome them. People who belong to family background having history of diabetes are 25% more prone to develop diabetes. Diet, is also a major factor responsible for causing diabetes. Obesity is also one of the major factors causing diabetes. Virus **infections** can also lead to diabetes. Age is the commonest predisposing factor for diabetes Emotional stress, also lead to diabetes.

infection[ɪnˈfekʃən]
n. 感染;传染

3. Diabetes **diagnosis**

There are several tests to diagnose diabetes mellitus. Some of the tests can be carried out at the household level, while there are some sophisticated ones that are carried out only in clinics. Whatever be the case, the diagnostic tests of diabetes mellitus are quite efficient in finding out the onset of the disease.

diagnosis [ˌdaɪəgˈnəʊsɪs]
n. 诊断

4. Regulation of glucose **transporter translocator** in health and diabetes

Fatigue is cause by dehydration from increased urination and the body's inability to function properly, since it's less able to use sugar for energy needs. Although research not been proven if this is entirely true, doctor believe frequent infections and sores may be caused by the high levels of blood sugar impairing the body's natural healing process and the ability to fight infections. The other symptoms include recurring skin, **gum**, or **bladder** infections. Diabetes may weaken the body's ability to fight **germs**, which increases the risk of infection in the gums and in the bones that hold the patient teeth in place. The gums may pull away from the teeth, the teeth may become loose,

transporter [trænˈspɔːtə]
n. 运输者
translocator [trænzləˈkeɪtə]
n. 转运蛋白;转位分子

gum [gʌm]
n. 牙龈
bladder [ˈblædə]
n. 膀胱;囊状物
germ [dʒɜːm]
n. 细菌

or the patient may develop sores or pockets of pus in your gums, especially if you have a gum infection before diabetes develops.

The exact cause of diabetes has not been identified, but it is caused by multiple factors and **hereditary** components. Individuals, who have a parent or sibling with type II diabetes, have a 10 percent to 15 percent chance of developing the disease. The risk is much higher if the sibling is an identical twin. An Inactive lifestyle or poor diet may act as a trigger for someone with a genetic tendency towards type II diabetes. The factor can vary from person to person. Some well known factors are being overweight and having defective beta cells. Risk factors include having high blood pressure of 140/90 mm Hg or higher, race and **high-density lipoprotein** (HDL) cholesterol less than 35 milligrams per deciliter (mg/dL). One of the most important risk factor of type II diabetes is processed foods, especially in many North American, European, and Asian countries. For example, the rates of type II diabetes in Spain have increased magnificently over the past 20 years, partly due to the Spaniards moving away from the traditional Mediterranean diet. Africans, Hispanics, Latinos, Pacific Islanders, and Asians have a high risk of developing type II diabetes. Other potential causes of the illness included chronic stress, low birth weight, and **gene mutations.** If type II is suspected, the patient should receive a diagnosis immediately.

As symptoms develop gradually, the patient is advised to receive a diagnosis. The doctor examines the person's medical history and does a physical examination. The common tests done are

---

hereditary [hɪˈredɪtərɪ]
adj. 遗传的；世袭的；世代相传的
n. 遗传类

high-density lipoprotein [haɪ-ˈdensɪtɪ, ˈlɪpəprəutin]
n. 高密度脂蛋白

gene mutation [dʒiːn mjuːˈteɪʃən]
n. 基因突变

a random blood test and a **fasting blood test**. The random blood test is usually done when the patient receives physical. If the random blood test reveals blood sugar level greater or equal to 200 mg/dL, then they are suspected of type II diabetes. In a fasting blood test, the patient fasts for eight hours and then is tested for amount of glucose. Both testes are later confirmed for type II diabetes with the **glucose tolerance test**. In the glucose tolerance test, the patient cannot eat or drink anything after midnight before the test. For the test, the patient will be asked to drink a liquid containing a certain amount of glucose. The victim's blood will be taken before drinking the liquid, and again every 30 to 60 minutes after the patient drink the solution. The test could take up to three hours. In a patient without diabetes, the drink will cause the glucose level to rise quickly and then fall as the body metabolizes the glucose. In a patient with the illness, the blood sugar level will increase greatly, but instead of gradually falling, the glucose level remains elevated. After testing is completed, different ways of treatments will be used.

After a patient is diagnosed with type II diabetes, he/she will take the next series of steps to maintain blood sugar at a level closest to normal. Patients are supposed to test their blood sugar level every day. One of the key elements of home blood glucose measurement (HBGM) is glucose meter. It is a medical device that determines the approximate concentration of glucose in the blood. A small drop of blood, obtained by pricking the skin with a **lancet**, is placed on a disposable test strip that the meter reads and uses to calculate the blood glucose level. It is one of most inexpensive

fasting blood test
['fɑːstɪŋ blʌd test]
n. 空腹验血

glucose tolerance test
['gluːkəʊs 'tɒlərəns test]
n. 葡萄糖耐量测试

lancet ['lɑːnsɪt]
n. 小刀，柳叶刀

ways to test glucose level because one test strip is about 35 cents. Although it is not used as often as type I diabetes victims, but it can be critical to maintaining a healthy glucose level. Most people with type II diabetes need to check their blood sugar once or twice a day.

Type II diabetes also can be managed with physical exercises and dietary modification. Regular exercise is important for everyone, but especially if you have diabetes. Regular **aerobic** exercise lowers your blood sugar level without **medication** and helps burn excess calories and fat so you can manage your weight. Even with physical excises and dietary modification, medication may be still needed.

Some drugs are used to treat type II diabetes. One common type of injectable medication is Incretin. Incretin is used to stimulate the pancreas to produce insulin and suppress appetite, therefore leads to weight loss. Oral drug is taken by mouth to lower blood sugar level. Metformin, also known as Biguanides, is an oral medication taken three times a day that is used to reduce body production of glucose and help muscle, fat cells and liver absorb more glucose from the bloodstream. This lowers blood sugar levels. Amylin Analogues is used to replace protein that is normally produced by the pancreas. It is taken before each meal. α glucosidase inhibitors such as **acarbose**, decrease the absorption of carbohydrates from the digestive tract to lower after-meal glucose levels. **Bariatric** (weight loss) surgery may be considered for very overweight patients who are not well managed with diet and medications.

aerobic [eəˈrəʊbɪk]
adj. 需氧的
medication [medɪˈkeɪʃən]
n. 药物；药物治疗；药物处理

acarbose [ˈækɑrˌbəʊs]
n. 阿卡波糖
bariatric [ˌbærɪˈætrɪks]
adj. 肥胖症治疗学的

Long-term complications from high blood sugar include increased risk of heart attacks, **cardiovascular** diseases, strokes, **amputation**s, and **kidney failures**. In addition, diabetes affects the body's **immune** system. This decreases the body's ability to fight infection. Small infections can quickly get worse and cause the death of skin and other tissues. Type II diabetes victims are more likely to have foot problems. Diabetes can damage nerves, which means you may not feel an injury to the foot until a large sore or infection develops. About 60% to 70% of people with the illness have mild to severe forms of **nervous system** damage. Diabetes can also damage **blood vessel**, which may lead to other serious injuries.

Common names for type II diabetes are non-insulin dependent diabetes and adult on-set diabetes. Type II diabetes affects many **geographical** locations around the world. The World Health Organization (WHO) estimates 285 million people have type II diabetes that is about 4.6 percent of the world's adult population in 2010. The number of patients is expected to reach 438 million by 2030 that will be 8.5 percent of the world's population. India has the most type II diabetes victims with 50.8 million. In 2004, an estimated 50.8 million people die from high blood pressure. In the United States, it is the 6th leading cause of death in 2008. The United States government spent an estimated 174 billion dollars directly and indirectly on diabetes. Diabetes is the leading cause of new cases of blindness among 20 to 75 years olds in the United States. It is also the leading cause of kidney failure in the United States, accounting for 44 percent of the new causes.

cardiovascular [kɑːdɪəʊ'væskjʊlə]
adj. 心血管的
amputation [æmpjʊ'teɪʃən]
n. 截肢；切断（术）
kidney failure ['kɪdni 'feiljə]
n. 肾功能衰竭
immune [ɪ'mjuːn]
adj. 免疫的

nervous system ['nəːvəs 'sistəm]
n. 神经系统
blood vessel [blʌd 'vesəl]
n. 血管
geographical [dʒɪə'græfɪkəl]
adj. 地理的；地理学的

The first symptoms were documented by Hesy-Ra, an Egyptian **physician**. The symptoms were frequent urination and it was called the mysterious disease. Aretaeus, a Greek physician, described the destructive nature of the affliction which he named "diabetes" from the Greek word for siphon in the first century AD. Ancient Indians call diabetes "sweet urine disease," and observe if the ants are attracted to urine as a test for diabetes. Based on animal research, Dr. Joseph von Mering and Oskar Minkowski discovered that the pancreas plays a role in diabetes in 1900. In 1936, Sir Harold Percival Himsworth successfully distinguished type I diabetes from type II diabetes. Many people know someone with type II diabetes, but they do not realize what it is. Understanding type II diabetes and how to control it are important. There is currently no cure for type II diabetes and scientists are still working on it.	physician [fɪ'zɪʃən] *n.* 内科医师

参考译文

1. 什么是糖尿病？

　　糖尿病史一种由胰岛素生成不足或身体无法对系统中生成的胰岛素做出反应而引起的代谢疾病。这种疾病可以分为两种不同的类型：Ⅰ型和Ⅱ型。Ⅰ型糖尿病是有胰腺中胰岛β细胞的缺失引起的。另一种类型的糖尿病是Ⅱ型糖尿病。Ⅱ型糖尿病通常指的是对人体胰岛素的抵抗。这主要归因于某些组织中胰岛素受体的损失应该调节胰岛素进入人体的细胞。

　　这个特定类型的糖尿病是折磨的大多数的病例报道中最常见的一种。Ⅱ型糖尿病通常会导致高血糖。所有类型的糖尿病都有其他疾病的某些不同的症状，它们也都有不同的并发症。1型糖尿病是指没有胰岛素代谢葡萄糖和导致血液中葡萄糖水平偏低的情况，Ⅱ型糖尿病指的是血液中有异常高的葡萄糖的情况。

2. 导致糖尿病的原因是什么？

　　糖尿病的原因并不十分清楚。仍然是一个未解之谜。医学科学家无法解释为什么有些人患有糖尿病，而其他人没有。有一些因素可能会导致糖尿病。这些危险因素，我们每个人都要试图去克服。具有家族遗传病史背景的人比其他人患糖尿病的概率增加25%。饮食也是引发糖尿病的一个主要因素。肥胖症是引发糖尿病的另一个主要因素。病毒感染也会导致糖尿病。年龄是最常见的诱发因素，另外，精神压力也会导致糖尿病。

3. 糖尿病诊断

有几个测试方法可以用来诊断糖尿病。一些测试可以在家庭层面进行，但是仍有一些复杂的测试只能在诊所进行。尽管如此，糖尿病诊断测试能非常有效的发现疾病的发作。

4. 葡萄糖转运蛋白在健康状态和糖尿病中的调节作用

由于用来产能的糖类供应不足，会导致排尿引发的脱水和身体功能无法正常进行。尽管研究没有证明是否完全正确，医生认为频繁的感染和溃疡可能是由于高水平的血糖损害人体的自然愈合过程和抵抗感染的能力。其他症状包括反复出现皮肤、牙龈或膀胱感染。糖尿病可能会削弱身体抵抗病菌的能力，这就增加了病人牙齿在牙龈和骨骼感染的风险。牙龈可能脱离牙齿，牙齿变松动，或者患者可能出现溃疡或在牙龈化脓，特别是如果你在患糖尿病前有一个牙龈感染。

糖尿病的确切原因尚未确定，但它是由多种因素和遗传原因共同引起的。如果一个人的父母或兄弟姐妹患有 II 型糖尿病，那么他有 10%~15% 的机会也患该病。如果患病的亲人是同卵双胞胎，那么患病的风险还要高得多。不活跃的生活方式和不良的饮食习惯可能触发遗传 II 型糖尿病的倾向。因素可能因人而异。影响较大的因素主要是肥胖和 β 细胞缺陷。风险因素包括有较高的血压 140/90mmHg 或更高，种族和高密度脂蛋白（HDL）胆固醇少于 35mg/dL。II 型糖尿病的最重要的危险因素之一是加工食品，尤其是在许多北美、欧洲和亚洲国家。例如，西班牙的 II 型糖尿病在过去 20 年里大幅度增加，部分原因是西班牙人远离传统的地中海饮食。非洲人、西班牙人、拉丁美洲人、太平洋岛民和亚洲人有很高的患 II 型糖尿病的风险。其他潜在原因的疾病包括慢性压力，低出生体重和基因突变。如果怀疑 II 型，病人应该立即得到诊断。

随着症状的逐渐发展，建议病人应接受诊断。医生对患者的病史和体格分别进行检查。常见的检测是一个随机的血液测试和空腹验血。随机血液测试通常在体格测试中进行。如果随机的血液测试表明血糖水平大于或等于 200mg/dL，他们就疑似患有 II 型糖尿病。空腹验血的病人要禁食 8h，然后检测葡萄糖。这两项测试和后来的葡萄糖耐量试验都是用来证实 II 型糖尿病的。葡萄糖耐量试验中，病人在测试前的午夜不能吃或喝任何东西。对于测试，病人将被要求喝含有一定量葡萄糖的液体。病人在喝液体之前要进行采血，喝液体后的每 30~60min 再进行采血化验。这项测试持续 3h。在一个没有糖尿病的患者体内，这种饮料会导致血糖水平迅速上升，然后随着葡萄糖在身体内代谢而下降。疾病患者，血糖水平会大大增加，而不是逐渐下降，葡萄糖水平仍较高。测试完成后，将使用不同的治疗方式。

病人被诊断为 II 型糖尿病后，他/她接下来将维持血糖水平接近正常。病人应该每天测试血糖水平。家用血糖测量的关键元素之一（HBGM）是血糖仪。它是一个能测定近似血液中葡萄糖的浓度医疗设备。用柳叶刀刺破皮肤获得一小滴血，滴在一次性测试条上，其长度大小能够读取并用来计算血糖水平。它是测试血糖水平最便宜的方法，因为一个测试条大约是 35 美分。尽管它不经常被用于 I 型糖尿病患者，但是它是保持健康的血糖水平的关键。大多数的 II 型糖尿病患者需要检查血糖一天一次或两次。

II 型糖尿病也可以通过体育锻炼和饮食调节。定期锻炼对每个人来说都很重要，特别是对于糖尿病的病人。规律地做有氧运动在没有药物治疗的情况下也能降低血糖水平，并帮助燃烧多余的热量和脂肪，这样你就可以管理你的体重。即使身体运动与饮食调节，某些情况下，也仍然需要药物辅助治疗。

有些药物常用于治疗 II 型糖尿病。胰岛素是其中一个常见的注射药物。胰岛素可以被用来刺激胰腺产生胰岛素和抑制食欲，因此会导致体重减轻。口服药物也可以降低血糖水平。二甲双胍，也被称为双胍类药物，属于一天三次的口服药物，能够减少身体生产葡萄糖和帮助肌肉，脂肪细胞和肝脏从血液里吸收更多的葡萄糖。这类药物也可降低血糖水平。淀粉类似物可被用来替代有胰腺产生的蛋白质。在每次餐前服用。α-葡萄糖苷酶抑制剂如阿卡波糖，能够减少碳水化合物从消化道的吸收而降低饭后血糖水平。减肥（减肥）才可以考虑手术非常超重患者不能很好地与饮食和药物管理。

长期高血糖的并发症包括心脏病发作的风险增加、心血管疾病、中风、截肢和肾脏的衰竭。此外，糖尿病会影响人体的免疫系统，并降低身体抵抗感染的能力。小感染可以迅速恶化并导致皮肤和其他组织的死亡。II 型糖尿病患者更容易在脚部发病。因为糖尿病会损害神经，这意味着你可能感觉不到脚受伤，直到疼痛或感染的发展。60%~70% 的病人有轻微到严重的神经系统损伤。糖尿病也会损害血管，这可能会导致其他严重损伤。

II 型糖尿病的常用名字是非胰岛素依赖型糖尿病和成人糖尿病。世界各地的许多地理位置均受到 II 型糖尿病的影响。世界卫生组织（WHO）估计有 2.85 亿人罹患二型糖尿病，占 2010 年世界成年人口的 4.6%。到 2030 年，病人的数量有望达到 4.38 亿，届时将占到 8.5% 的世界人口。印度的 II 型糖尿病患者最多，已达 5.08 亿人。在 2004 年，大约 5.08 亿人死于高血压。2008 年，这是美国第六大死亡的主要原因。美国政府花了大约 1740 亿美元直接或间接用来应对糖尿病。糖尿病是美国 20~75 岁的新病例失明的主要原因。在美国，这也是肾衰竭的主要原因，占新病例的 44%。

第一个记录糖尿病症状的是埃及医生 Hesy-Ra。由于糖尿病伴有尿频的症状，因而被称为神秘的疾病。希腊医生 Aretaeus 在公元 1 世纪对该病的破坏性进行描述，命名为"糖尿病"。古印度人称糖尿病为"甜尿病"，并观察蚂蚁是否被尿液吸引作为糖尿病的测试。基于动物研究，Joseph von Mering 博士和 Oskar Minkowski 在 1900 年发现，胰腺在糖尿病中发挥作用。1936 年，Harold Percival Himsworth 爵士成功地区分了 II 型糖尿病 I 型糖尿病。

许多人知道有人患有 II 型糖尿病，但他们不知道它是什么。理解 II 型糖尿病和如何控制它是很重要的。目前还没有治疗 II 型糖尿病的方法，科学家们仍在研究。

Further Reading
Definition and Description of Diabetes Mellitus

Diabetes mellitus is a group of metabolic diseases characterized by **hyperglycemia** resulting from defects in insulin secretion, insulin action, or both. The chronic hyperglycemia of diabetes is associated with long-term damage, dysfunction, and failure of various organs, especially the eyes, **kidneys**, nerves, heart, and blood vessels.

Several pathogenic processes are involved in the development of diabetes. These range from autoimmune destruction of the cells of the pancreas with consequent insulin deficiency to abnormalities that result in resistance to insulin action. The basis of the abnormalities in carbohydrate, fat, and protein metabolism in diabetes is deficient action of insulin on target tissues. Deficient insulin action results from inadequate insulin secretion and/or diminished tissue responses to insulin at one or more points in the complex pathways of hormone action. Impairment of insulin secretion and defects in insulin action frequently coexist in the same patient, and it is often unclear which **abnormality**, if either alone, is the primary cause of the hyperglycemia.

Symptoms of marked hyperglycemia include polyuria, polydipsia, weight loss, sometimes with polyphagia. Impairment of growth and susceptibility to certain infections may also accompany chronic hyperglycemia. Acute, life-threatening consequences of uncontrolled diabetes are hyperglycemia with keto acidosis or the totic hyperosmolar syndrome.

Long-term complications of diabetes include retinopathy with potential loss of vision; nephropathy leading to renal failure; peripheral neuropathy with risk of foot ulcers, amputations, and Charcot joints; and autonomic neuropathy causing gastro intestinal, genitourinary, and cardiovascular symptoms and sexual dysfunction. Patients with diabetes have an increased incidence of atherosclerotic cardiovascular, peripheral arterial, and cerebrovascular disease. Hypertension and abnormalities of lipoprotein metabolism are often found in people with diabetes. The vast majority of cases of diabetes fall into two broad genetic categories (discussed in greater detail below).

In one category, type 1 diabetes, the cause is an absolute deficiency of insulin secretion. Individuals at increased risk of developing this type of diabetes can often be identified by serological evidence of an auto immune pathologic process occurring in the pancreatic islets and by genetic markers. In the other, much more prevalent category, type 2 diabetes, the cause is a combination of resistance to insulin action and an inadequate compensatory insulin secretory response. In the latter category, a degree of hyperglycemia sufficient to cause pathologic and functional changes in various target tissues, but without clinical symptoms, may be present for a long period of time before diabetes is detected. During this asymptomatic period, it is possible to demonstrate an abnormality in carbohydrate metabolism by measurement of plasma glucose in the fasting state or after a challenge with an oral glucose load.

New Words

hyperglycemia [ˌhəipəɡləisiːmiə] *n.* 高血糖症

kidney ['kidni] *n.* 肾脏

abnormality [ˌæbnɔr'mæləti] *n.* 异常；畸形

参考文献

[1] Norris S L, Zhang X, et al. Long-term non-pharmacological weight loss interventions for adults with type 2 diabetes mellitus. Sao Paulo Medical Journal, 2016, 134(2):85-86.

[2] Switzer N J, Prasad S, et al. Sleeve Gastrectomy and Type 2 Diabetes Mellitus: a Systematic Review of Long-Term Outcomes. Obesity Surgery, 2016, 26(7):1-6.

[3] Avogaro A, Fadini G P, et al. Endothelial dysfunction in type 2 diabetes mellitus. Indian Journal of Clinical Biochemistry, 2016, 16(4):S39-S45.

[4] Aronson D, Edelman E R. Coronary Artery Disease and Diabetes Mellitus. Heart Failure Clinics, 2016, 12(1):117-133.

[5] Switzer N J, Prasad S, et al. Sleeve Gastrectomy and Type 2 Diabetes Mellitus: a Systematic Review of Long-Term Outcomes. Obesity Surgery, 2016, 26(7):1-6.

Unit 3　Microbiology　微生物

Chapter 1　Introduction of Microbiology　微生物的介绍

One can't overemphasize the importance of microbiology. Society benefits from **microorganisms** in many ways. They are necessary for the production of bread, cheese, beer, **antibiotics**, **vaccines**, vitamins, enzymes, and many other important products. Indeed, modern biotechnology rests upon a microbiological foundation. Microorganisms are indispensable components of our **ecosystem**. They make possible the cycles of carbon, oxygen, nitrogen, and sulfur that take place in **terrestrial** and **aquatic** systems. They also are a source of nutrient sat the base of all **ecological** food chains and webs.

1. The discovery of microorganisms

Even before microorganisms were seen, some investigators believe their existence and responsibility for disease. Among others, the Roman **philosopher** Lucretius (about 98—55 s.c.) and the **physician** Girolamo Fracastoro (1478—1553) suggested that disease was caused by invisible living **creatures**. The earliest microscopic observations appear to have been made between 1625 and 1630 on bees and **weevils** by the Italian Francesco Stelluti, using a microscope probably supplied by Galileo. However, the first person to observe and describe microorganisms accurately was the amateur microscopist Antony van Leeuwenhoek (1632—1723) of Delft, Holland. Leeuwenhoek earned his living as a draper and haberdasher (dealer

New Words and Expressions
microorganism [maɪkrəʊˈɔːɡənɪzəm]
n. 微生物；微小动植物
antibiotic [ˌæntɪbaɪˈɒtɪk]
n. 抗生素
vaccine [ˈvæksiːn]
n. 疫苗；牛痘苗
ecosystem [ˈiːkəʊsɪstəm]
n. 生态系统
terrestrial [təˈrestrɪəl]
adj. 地球的；陆地的，[生物] 陆生的
aquatic [əˈkwætɪk]
adj. 水生的；水栖的
ecological [iːkəˈlɒdʒɪkəl]
adj. 生态的，生态学的
philosopher [fɪˈlɒsəfə]
n. 哲学家；哲人
physician [fɪˈzɪʃən]
n. 医师；内科医师
creature [ˈkriːtʃə]
n. 动物，生物；人；创造物
weevil [ˈwiːvəl]
n. 象鼻虫

in men's clothing and accessories), but spent much of his spare time constructing simple **microscopes** composed of double **convex** glass held between two silver plates. His microscopes could magnify around 50 to 300 times, and he may have illuminated his liquid **specimens** by placing them between two pieces of glass and shining light on them at a 450 angle to the specimen plane. This would have provided a form of **dark-field illumination** and made bacteria clearly visible. Beginning in 1673 Leeuwen-hoek sent detailed letters describing his discoveries to the Royal Society of London. It is clear from his descriptions that he saw both bacteria and **protozoa**.

2. The relationship between microorganisms and diseases

Although Fracastoro and a few others had suggested that invisible organisms produced disease, most believed that disease was due to causes such as supernatural forces, poisonous vapors called **miasmas**, and imbalances between the four humors thought to be present in the body. The idea that an imbalance between the four humors (blood, **phlegm**, yellow bile "**choler**", and black bile "melancholy") led to disease had been widely accepted since the time of the Greek physician Galen. Support for the **germ** theory of disease began to accumulate in the early nineteenth century. Agostino Bassi (1773—1856) first showed a microorganism could cause disease when he demonstrated in 1835 that a **silkworm** disease was due to a **fungal** infection. He also suggested that many diseases were due to microbial infections. In 1845 M. J. Berkeley proved that

microscope ['maɪkrəˌskəup]
n. 显微镜
convex ['kɒnveks]
n. 凸面体；凸状
adj. 凸面的；凸圆的
specimen ['spesɪmɪn]
n. 样品，样本；标本
dark-field illumination
['dɑːkfˈiːld ɪˌluːmɪ'neɪʃn]
n. (显微镜的)暗(视)场照明(法)

protozoa [ˌprəutəu'zəuə]
n. [无脊椎]原生动物

miasma [mɪ'æzmə]
n. 瘴气；臭气；不良影响

phlegm [flem]
n. 痰；黏液；黏液质
choler ['kɒlə] *n.* 胆汁；愤怒
germ [dʒɜːm]
n. [植]胚芽，萌芽；细菌

silkworm ['sɪlkwɜːm]
n. 蚕，桑蚕
fungal ['fʌŋɡəl] *adj.* 真菌的

the great **Potato Blight** of Ireland was caused by a fungus. Following his successes with the study of fermentation, Pasteur was asked by the French government to investigate the **pebrine** disease of silkworms that was disrupting the silk industry. After several years of work, he showed that the disease was due to a protozoan parasite. The disease was controlled by raising **caterpillars** from eggs produced by healthy moths.

Indirect evidence that microorganisms were agents of human disease came from the work of the English surgeon Joseph Lister (1827—1912) on the prevention of wound infections. Lister impressed with Pasteur's studies on the involvement of microorganisms in fermentation and **putrefaction**, developed a system of **antiseptic surgery** designed to prevent microorganisms from entering wounds. Instruments were heat **sterilized**, and **phenol** was used on surgical dressings and at times **sprayed** over the surgical area. The approach was remarkably successful and transformed surgery after Lister published his findings in 1867. It also provided strong indirect evidence for the role of microorganisms in disease because phenol, which killed bacteria, also prevented wound infections.

The first direct demonstration of the role of bacteria in causing disease came from the study of **anthrax** by the German physician Robert Koch (1843—1910). Koch used the **criteria** proposed by his former teacher, Jacob Henle (1809—1885), to establish the relationship between ***Bacillus anthracis*** and anthrax, and published his findings in 1876 briefly discusses the scientific method). Koch injected healthy mice with material from diseased animals, and the mice became ill. After transferring

Potato Blight [pə'teitəu blaɪt]
n. 马铃薯枯萎病

pebrine [peb'rin]
n. 微粒子病

caterpillar ['kætəpɪlə]
n. [无脊椎] 毛虫

putrefaction [ˌpjuːtrɪ'fækʃən]
n. 腐败；腐败物
antiseptic [ˌænti'septɪk]
adj. 防腐的，抗菌的；非常整洁的
surgery ['sɜːdʒərɪ]
n. 外科；外科手术；手术室；诊疗
sterilized ['sterəlaɪzd]
v. 杀菌；消除；冻结
adj. 无菌的；已消过毒的
phenol ['fiːnɒl]
n. 苯酚
spray [spreɪ]
n. 喷雾；喷雾器；水沫
anthrax ['ænθræks]
n. 炭疽，炭疽热
criteria [kraɪ'tɪərɪə]
n. 标准，条件
Bacillus anthracis
[bə'sɪləs sænθ'reɪsɪz]
n. 炭疽杆菌，炭疽芽孢杆菌

anthrax by inoculation through a series of 20 mice, he **incubated** a piece of **spleen** containing the anthrax bacillus in beef **serum**. The **bacilli** grew, reproduced, and produced spores. When the isolated bacilli or spores were injected into mice, anthrax developed. His criteria for proving the causal relationship between a microorganism and a specific disease are known as Koch's postulates and can be summarized as follows:

（1）The microorganism must be present in every case of the disease but absent from healthy organisms.

（2）The suspected microorganism must be isolated and grown in a pure culture.

（3）The same disease must result when the isolated microorganism is inoculated into a healthy host.

（4）The same microorganism must be isolated again from the diseased host.

Although Koch used the general approach described in the postulates during his anthrax studies, he did not outline them fully until his 1884 publication on the cause of **tuberculosis**.

Koch's proof that Bacillus anthracis caused anthrax was independently confirmed by Pasteur and his coworkers. They discovered that after burial of dead animals, anthrax spores survived and were brought to the surface by **earthworms**. Healthy animals then ingested the spores and became ill.

3. The development of techniques for studying microbial **pathogens**

During Koch's studies on bacterial diseases, it became necessary to isolate suspected bacterial pathogens. At first he cultured bacteria on the

incubate [ˈɪŋkjʊbeɪt]
vt. 孵化；培养；温育
spleen [spliːn]
n. 脾脏
serum [ˈsɪərəm]
n. 血清；浆液；免疫血清；乳清；树液
bacilli [bəˈsɪlaɪ]
n. 杆菌

tuberculosis [tjʊˌbɜːkjʊˈləʊsɪs]
n. 肺结核；结核病

earthworm [ˈɜːθwɜːm]
n. 蚯蚓

pathogen [ˈpæθədʒən]
n. 病原体；病菌

sterile surfaces of cut, boiled potatoes. This was unsatisfactory because bacteria would not always grow well on potatoes. He then tried to **solidify** regular liquid media by adding **gelatin**. Separate bacterial colonies developed after the surface had been streaked with a bacterial sample. The sample could also be mixed with liquefied gelatin medium. When the gelatin medium hardened, individual bacteria produced separate colonies. Despite its advantages gelatin was not an ideal solidifying agent because it was digested by many bacteria and melted when the temperature rose above 28 ℃. A better alternative was provided by Fannie Eilshemius Hesse, the wife of Walther Hesse, one of Koch's assistants. She suggested the use of **agar** as a solidifying agent-she had been using it successfully to make **jellies** for some time. Agar was not attacked by most bacteria and did not melt until reaching a temperature of 100 ℃. One of Koch's assistants, Richard Petri, developed the dish (plate), a container for solid culture media. These developments made possible the isolation of pure cultures that contained only one type of bacterium, and directly stimulated progress in all areas of **bacteriology**.

Koch also developed media suitable for growing bacteria isolated from the body. Because of their similarity to body fluids, meat extracts and protein digests were used as nutrient sources. The result was the development of nutrient broth and nutrient agar, media that are still in wide use today.

By 1882 Koch had used these techniques to isolate the bacillus that caused tuberculosis. There followed a golden age of about 30 to 40 years in which most of the major bacterial pathogens were isolated.

The discovery of viruses and their role in disease

solidify [sə'lɪdɪfaɪ]
vt. 固结；凝固
gelatin ['dʒelətɪn]
n. 明胶；动物胶；胶制品

agar ['eɪgɑː]
n. 琼脂（一种植物胶）
jellies ['dʒelɪz]
n. 果胶；凝胶剂

bacteriology [bækˌtɪəri'ɒlədʒi]
n. 细菌学

was made possible when Charles Chamberland (1851—1908), one of Pastear's associates, constructed a **porcelain** bacterial filter in 1884. The first viral pathogen to be studied was the tobacco **mosaic** disease virus.

4. Members of the microbial world

The early description of organisms as either plants or animals clearly is too simplified, and for many years biologists have divided organisms into five kingdoms: the **Monera**, **Protista**, Fungi, animal, and **Plantae**. Microbiologists study primarily members of the first three kingdoms. Although they are not included in the five kingdoms, viruses are also studied by microbiologists.

In the last few decades great progress has been made in three areas that profoundly affect microbial classification. First, much has been learned about the detailed structure of microbial cells from the use of electron microscopy. Second, microbiologists have determined the biochemical and **physiological** characteristics of many different microorganisms. Third, the sequences of nucleic acids and proteins from a wide variety of organisms have been compared. It is now clear that there are two quite different groups of procaryotic organisms: Bacteria and **Archaea**. Furthermore, the protists are so diverse that it may be necessary to divide the kingdom Protista into three or more kingdoms. Thus many **taxonomists** have concluded that the five kingdom system is too simple and have proposed a variety of alternatives. The differences between Bacteria, Archaea, and the eucaryotes seem so great that many microbiologists have proposed thatorganisms should be divided among three domains: Bacteria (the true bacteria or eubacteria), Archaeal, and Eucarya (all eucaryotic organisms).

porcelain [ˈpɔːsəlɪn]
n. 瓷；瓷器
mosaic [məʊˈzeɪɪk]
adj. 花叶的

monera [məˈnɪərə]
n. 原核生物界
protista [ˈprəʊtistə]
n. 原生生物界
plantae [ˈplænˌtiː]
n. 植物界

physiological [ˌfɪzɪəˈlɒdʒɪkəl]
adj. 生理学的，生理的

archaea [aˈkiə]
n. 古生菌

taxonomist [tækˈsɔnəmist]
n. 分类学者

参考译文

人们怎样强调微生物学的重要性也是不过分的，社会在许多方面从微生物中获得益处，面包、乳酪、啤酒、抗生素、疫苗、维生素、酶和许多其他重要产品的生产都需要微生物。实际上，现代生物技术是建立在微生物学的基础上。微生物是我们的生态系统不可缺少的组成成员，它们使陆地和水生系统中碳、氧、氮和硫的循环成为可能，它们也是所有生态食物链和食物网的根本营养来源。

1. 微生物的发现

在微生物被看见之前，一些研究者就已猜想到它们的存在，并且认为它们是引起疾病的原因。其中，古罗马哲学家 Lucretius（公元前 98—前 55 年），内科医生 Girolamo Fracastoro（1478—1553 年）提出：疾病是由看不见的活动生物引起的。意大利人 Rrancesco Stelluti 在 1625—1630 年，对蜜蜂和象鼻虫最早进行了显微镜观察，他们使用的显微镜可能是由 Galil 提供的。然而，第一个观察和描述微生物的人，准确地说应是荷兰德尔夫特人安东•范•列文虎克（1632—1723 年），他是一位业余显微镜爱好者。列文虎克虽然是一位布商和男子服饰用品商，但是却用了大量的业余时间制造简单的显微镜，这种显微镜是将双凸玻璃镜安放在两个银片之间而构成。他的显微镜能够放大约 50 倍，他将液体样品保存，放在两块玻璃之间，光线以 45° 角照射在样品表面，可以照亮液体样品，这样就会使暗视野照亮，而能清楚地看见细菌。从 1673 年开始，列文虎克给伦敦皇家学会寄去的所有信中，都详细地描述了他的发现。从他的叙述中，能够清楚地知道列文虎克既看见了细菌又看见了原生动物。

2. 微生物与疾病之间的关系

虽然 Lracastor 和其他学者已提出：看不见的有机体能引起疾病，但大多数人仍相信：疾病是由超自然力量、称为瘴气的毒气和存在体内的 4 种液体之间不平衡所引起。自从希腊医生 Galen 时代以来，人们已广泛地接受了疾病是由于 4 种液体（血液、黏液、黄色胆汁"怒气症"和黑胆汁"忧郁症"）之间不平衡而引起的观念。19 世纪初期，对病原菌理论的支持开始积聚。当 Agostino Bassi（1773—1856 年）1835 年证实了一种蚕病是由真菌感染引起时，他就首先指出微生物能够引起疾病，也提出许多疾病是由于微生物感染引起的。1845 年，M. J. Berkeley 证明爱尔兰严重的马铃薯疫病是由一种真菌引起。随着巴斯德发酵研究的成功，法国政府请求他调查研究破坏蚕丝工业的蚕微粒子，几年工作之后，巴斯德指出这种病是由一种寄生的原生动物引起，可通过提高从健康蚕蛾产的卵形成的幼虫的比率来控制这种病。

微生物是人类致病因子的间接证据来自于英国外科医生 Joseph Lister（1827—1912 年）关于创伤感染的预防。他受巴斯德关于微生物参与发酵和腐败的研究结果的启迪，建立了防止微生物进入创口的防腐外科手术系统，例如，加热灭菌外科器具，用石炭酸消毒，并时常喷雾石炭酸消毒手术的空间。1867年，Liste 发表了他的研究结果后，这种方法获得了显著的成功，并改造了外科手术。因为能杀死细菌的石炭酸也能防止创伤感染，所以这也为微生物在疾病中的作用提供了强有力的间接证据。

细菌引起疾病的第一个直接证据，来自于德国医生罗伯特•柯赫（1843—1910年）对炭疽的研究。柯赫采用了他的老师 Jacou Henle（1809—1885年）提议的准则，建立了炭疽芽孢杆菌（*Bacillus anthracis*）与炭疽之间的关系，并在1876年公布了他的发现。柯赫用来自有病动物的物质注射健康的老鼠，这些老鼠都发病。通过20只老鼠一系列的接种转移炭疽之后，他将一块含有炭疽芽孢杆菌的脾脏放在牛血清中培养，这种杆菌生长、繁殖和产生孢子，当用分离出的杆菌或孢子注射老鼠时，炭疽病就发生了。从而证明一种微生物与一种特定疾病之间的因果关系，被称为柯赫定律，该定律概括如下。

（1）病原微生物一定存在于一切患病个体中，而在健康的个体中不存在。
（2）一定能分离和纯培养所怀疑的病原微生物。
（3）当分离的病原微生物接种健康的宿主时，一定导致相同的疾病。
（4）相同的病原微生物一定再从这种发病的宿主中分离到。

尽管柯赫在他的炭疽研究中应用了该定律中所描述的一般方法，但一直到1884年他发表了结核病的病因，才充分地论述了该定律。

巴斯德和他的同事们也用自己的工作证实了柯赫关于炭疽芽孢杆菌引起炭疽病的论述。他们发现死亡的动物埋葬后，炭疽孢子仍是活的，并被蛆虫携带到地面，当这些孢子被健康动物摄取后，动物便发病。

3. 微生物病原体研究技术的发展

柯赫研究细菌病时，他需要分离被怀疑的细菌病原体。最初，他在煮过的马铃薯无菌切面上培养细菌，这种方法不能令人满意，因为细菌不会总是在马铃薯上长得好。后来他尝试在液体培养基中加入明胶制成的固态培养基，将细菌样品划线在培养基表面后，则长出分散开的细菌菌落，样品也可以与融化的明胶培养基混合，当明胶培养基凝固后，单个的细菌产生了分散开的菌落。尽管明胶有其优点，但它不是理想的凝固剂，因为许多细菌可以消化它，而且当温度升到约25℃时则被融化。Fannie Filshemius Hesse 提供了一个较好的代用品，她是柯赫的助手之一、Walther Hesse 的妻子。她建议利用琼脂作为凝固剂，她利用琼脂成功地制作果子冻已有一段时期。琼脂不被大多数细菌所利用，温度一直达到100℃也不融化。柯赫助手之一发明出了培养皿或平板，这是一种盛固体培养基的容器。这些进展不仅使分离纯培养物（只含一种类型的细菌）成为可能，而且直接促进了细菌学所有领域的进步，建立了细菌的分离和纯净培养技术。

柯赫也研制了适合从身体中分离出的细菌生长的培养基，由于肉提取物和蛋白质水解液与体液相似,所以用它们作为营养来源,结果研制出了今天还广泛使用的营养肉汤和营养琼脂培养基。

1882年,柯赫利用这些技术分离引起结核病的杆菌,随后是一个30～40年的黄金时代,在此年代中,分离出了大多数主要的病原菌。

1884年,巴斯德的同事之一Charles Chamberland (1851—1908年)构建了一个瓷的细菌过滤器,这使得发现病毒和它在疾病中的作用有了可能。被研究的第一个病毒病原体是烟草花叶病病毒。

4. 微生物界的成员

早期对有机体的描述是很清楚地简化为植物或动物。生物学家花费了许多年才将有机体分成五界:原核生物界、原生生物界、真菌界、动物界和植物界。微生物学家主要研究前三界的生物,虽然病毒没有包括在五界之中,但微生物学家也研究病毒。

在近几十年内,3个领域取得了很大进步,并深刻地影响了微生物的分类。首先,利用电子显微镜技术认识了微生物细胞的详细结构;第二,微生物学家已确定了许多不同微生物的生物化学和生理学特性;第三,对广泛的不同种类的有机体进行了核酸序列和蛋白质序列的比较。现已清楚,原核有机体存在两种完全不同的类群:细菌和古生菌,而且,原生生物也是多种多样,可能需要将原生生物界再分成3个或更多的界,所以许多分类学家断定五界系统是太简单,并已提出建议。细菌、古生菌和真核生物之间的差别显得如此之大,以致许多微生物学家已提议应将有机体划分成3个域:细菌(真正的细菌或真细菌)、古生菌和真核生物(所有的真核有机体)。我们在此将会采用该分类系统。

Further Reading
Industrial Microbiology and Microbial Ecology

Although Theodore Schwann and others had proposed in 1837 that yeast cells were responsible for the conversion of sugars to alcohol, a process they called alcoholic fermentation, the leading chemists of the time believed microorganisms were not involved. They were convinced that fermentation was due to a chemical instability that degraded the sugars to alcohol. Pasteur did not agree. It appears that early in his career Pasteur became interested in fermentation because of his research on the stereo chemistry of molecules. He believed that fermentations were carried out by living organisms and produced **asymmetric** products such as **amyl** alcohol that had optical activity. There was an intimate connection between molecular asymmetry, optical activity, and life. Then in 1856 M. Bigo, an industrialist in Lille, France, where Pasteur worked, requested Pasteur's assistance. His business produced alcohol from the fermentation of beet sugars, and the alcohol yields had recently declined and the product had become sour. Pasteur discovered that the fermentation was failing because the yeast normally responsible for alcohol formation had been replaced by microorganisms producing lactic acid rather than ethanol. In solving this practical problem, Pasteur demonstrated that all fermentation were due to the activities of specific yeasts and bacteria, and he published several papers on fermentation between 1857 and 1860. His success led to a study of wine diseases and the development of **pasteurization** to preserve wine during storage. Pasteur's studies on fermentation continued for almost 20 years. One of his most important discoveries was that some fermentative microorganisms were anaerobic and could live only in the absence of oxygen, whereas others were able to live either aerobically or anaerobically. A few of the early microbiologists chose to investigate the ecological role of microorganisms. In particular they studied microbial involvement in the carbon, nitrogen, and sulfur cycles taking place in soil and aquatic habitats. Two of the pioneers in this **endeavor** were Sergei N. Winogradsky (1856—1953) and Martinus W. Beljerlnck (1851—1931).

The Russian microbiologist Sergei N. Winogradsky made many contributions to soil microbiology. He discovered that soil bacteria could oxidize iron, sulfur, and ammonia to obtain energy, and that many bacteria could **incorporate** CO_2 into organic matter much like **photosynthetic** organisms do. Winogradsky also isolated anaerobic nitrogen-fixing soil bacteria and studied the decomposition of cellulose.

Martinus W. Beijerinck was one of the great general microbiologists who made fundamental contributions to microbial ecology and many other fields. He isolated the aerobic nitrogen-fixing bacterium **Azotobacter**; a root nodule bacterium also capable of fixing nitrogen; and sulfate-reducing bacteria. Beijerinck and Winogradsky developed the enrichment-culture technique and the use of selective media, which have been of such great importance in microbiology.

New Words
asymmetric [ˌæsɪ'metrɪk] *adj.* 不对称的；非对称的
amyl ['eɪmaɪl] *n.* 戊（烷）基
pasteurization [ˌpæstəraɪ'zeɪʃən] *n.* 加热杀菌法，巴斯德杀菌法
endeavor [ɪn'devə] *n.* 努力；尽力（等于 endeavour）
incorporate [ɪn'kɔːpəreɪt] *vt.* 包含，吸收；体现；把……合并
photosynthetic [ˌfəutəsin'θetik] *adj.* 光合的；光合作用的
azotobacter [ə'zəutəuˌbæktə] *n.* 固氮菌

参考文献

[1] Baker. J. J. R, Allen. G. E. prediction and implication in biology Reading, Mass.: Addison-Wesley.1968

[2] Ford. B. J. The earliest views. Sci. Am.1998, 278 (4):50—53.

[3] Drews. G. Ferdinand Cohn, a founder of modern microbiology. ASM News. 1999, 65 (8):547-53.
[4] Fredricks. D. N., Relman. D. A. Sequence-based identification of microbial pathogens: A reconsideration of Koch's postulates. Clin. Microbiol. Rev. 1996, 9 (1):18-33.

Chapter 2 Microorganism and Culture Media 微生物和培养基

1. The common nutrient requirements

Analysis of microbial cell composition shows that over 95% of cell dry weight is made up of a few major elements: carbon, oxygen, hydrogen, nitrogen, sulfur, phosphorus, **potassium,** calcium, **magnesium,** and **iron**. These are called **macroelements** or macronutrients because they are required by microorganisms in relatively large amounts. The first six (C, O, H, N, S, and P) are components of carbohydrates, lipids, proteins, and nucleic acids. The remaining four macroelements exist in the cell as canons and play a variety of roles. For example, potassium (K^+) is required for activity by a number of enzymes, including some of those involved in protein synthesis. Calcium (Ca^{2+}), among other functions, contributes to the heat resist ante of bacterial endospores. Magnesium (Mg^{2+}) serves as a cofactor for many enzymes, complexes with ATP, and stabilizes **ribosome**s and cell membranes. Iron (Fe^{2+} and Fe^{3+}) is a part of **cytochromes** and a cofactor for enzymes and electron-carrying proteins.

All organisms, including microorganisms, require several **micronutrients** or trace elements besides macroelements. The micronutrients manganese, **zinc, cobalt, molybdenum, nickel,** and copper-are needed by most cells. However, cells require such small amounts that contaminants in water, glassware, and regular media components often are adequate for growth. Therefore, it is very difficult to demonstrate a micronutrient requirement. In nature, micronutrients are ubiquitous and probably

New Words and Expressions

potassium [pəˈtæsɪəm] n. 钾
magnesium [mægˈniːzɪəm] n. 镁
iron [ˈaɪən] n. 铁
macroelement [ˈmækrəˈelimənt] n. 大量元素

ribosome [ˈraɪbəsəum] n. 核糖体
cytochrome [ˈsaɪtəukrəum] n. 细胞色素
micronutrient [maɪkrəuˈnjuːtrɪənt] n. 微量营养素
zinc [zɪŋk] n. 锌
cobalt [ˈkəubɔːlt] n. 钴；钴类颜料
molybdenum [məˈlɪbdənəm] n. 钼
nickel [ˈnɪkəl] n. 镍

do not usually limit growth. Micronutrients are normally a part of enzymes and cofactors, and they aid in the catalysis of reactions and maintenance of protein structure.

Besides the common macroelements and trace elements, microorganisms may have particular requirements that reflect the special nature of their morphology or environment. Diatoms need silicic acid (H_4SiO_4) to construct their beautiful cell walls of silica (SiO^{2-}). Although most bacteria do not require large amounts of sodium, many bacteria growing in saline lakes and oceans depend on the presence of high concentrations of sodium ion (Na^+).

Finally, it must be emphasized that microorganisms require a balanced mixture of nutrients. If an essential nutrient is in short supply, microbial growth will be limited regardless of the concentrations of other nutrients.

2. Nutritional types of microorganisms

In addition to the need for carbon, hydrogen, and oxygen, all organisms require sources of energy and electrons for growth to take place. Microorganisms can be grouped into nutritional classes based on how they satisfy all these requirements. We have already seen that microorganisms can be classified as either **heterotrophs** or **autotrophs** with respect to their preferred source of carbon. There are only two sources of energy available to organisms:

(1) light energy.

(2) the energy derived from oxidizing organic or inorganic molecules.

Phototrophs use light as their energy source; **chemotrophs** obtain energy from the oxidation of chemical compounds (either organic or inorganic).

heterotroph [ˈhetərəutrɔf]
n. 异养生物
autotrophs [ˈɔtəˌtrɔfs]
n. 自养生物

phototroph [ˈfəutəutrəuf]
n. 光养生物
chemotroph [kemətˈrɒf]
n. 化能营养生物

Microorganisms also have only two sources for bacteria cannot oxidize water but extract electrons from inorganic donors like hydrogen, hydrogen sulfide, and elemental sulfur. **Chemoorganotrophic** heterotrophs (often called chemoheterotrophs, chemoorganoheterotrophs, or even heterotrophs) use organic compounds as sources of energy, hydrogen, electrons, and carbon. Frequently the same organic nutrient will satisfy all these requirements. It should be noted that essentially all pathogenic microorganisms are chemoheterotrophs.

The other two nutritional classes have fewer microorganisms but often are very important ecologically. Some purple and green bacteria are **photosynthetic** and use organic matter as their **electron donor** and carbon source. These **photo organotrophic** heterotrophs (photo-organo-pheterotrophs) are common inhabitants of polluted lakes and streams. Some of these bacteria also can grow as photoautotrophs with molecular hydrogen as an electron donor. The fourth group, the **chemolithotrophic** autotrophs (chemolithoautotrophs), oxidizes reduced inorganic compounds such as iron, nitrogen, or sulfur molecules to derive both energy and electrons for **biosynthesis**. Carbon dioxide is the carbon source. A few chemolithotrophs can derive their carbon from organic sources and thus are **heterotrophic**. Chemolithotrophs contribute greatly to the chemical transformations of elements (e.g., the conversion of **ammonia** to nitrate or sulfur to sulfate) that continually occur in the ecosystem.

chemoorganotrophic [kemɔːgənət'rɒfɪk]
n. 有机化能营养的

photosynthetic [ˌfəʊtəsɪn'θetɪk]
adj. 光合的；光合作用的
electron donor [ɪ'lɛktrɑn 'dɒnə]
n. 电子供体
photo organotrophic ['fəʊtəʊ ɔːgənət'rɒfɪk]
n. 光能有机营养的
chemolithotrophic [keməʊlɪθəʊʃ'rɒfɪk]
adj. 无机化能营养的
biosynthesis [baɪəʊ'sɪnθɪsɪs]
n. 生物合成

heterotrophic [ˌhetərəʊ'trɒfɪk]
adj. 异氧的
ammonia [ə'məʊnɪə]
n. 氨

3. **Uptake** of nutrients by the cell

The first step in nutrient use is uptake of the required nutrients by the microbial cell. Uptake mechanisms must be specific-that is, the necessary substances, and not others, must be acquired. It does a cell no good to take in a substance that it cannot use. Since microorganisms often live in nutrient-poor habitats, they must be able to transport nutrients from **dilute solutions** into the cell against a concentration **gradient.** Finally, nutrient molecules must pass through a selectively permeable **plasma** membrane that will not permit the free passage of most substances. In view of the enormous variety of nutrients and the complexity of the task, it is not surprising that microorganisms make use of several different transport mechanisms. The most important of these are **facilitated diffusion**, **active transport**, and **group translocation**. Eucaryotic microorganisms do not appear to employ group translocation but take up nutrients by the process of endocytosis.

4. Culture media

Much of the study of microbiology depends on the ability to grow and maintain microorganisms in the laboratory, and this is possible only if suitable culture media are available. A culture medium is a solid or liquid preparation used to grow, transport, and store microorganisms. To be effective, the medium must contain all the nutrients the microorganism requires for growth. Specialized media are essential in the isolation and identification of microorganisms, the testing of antibiotic sensitivities, water and food analysis, industrial microbiology, and other activities. Although all microorganisms need sources of

uptake [ˈʌpteɪk]
n. 摄取；领会；举起

dilute solution
[daiˈljuːt səˈluːʃən]
n. 稀溶液

gradient [ˈgreɪdɪənt]
n. 梯度；坡度

plasma [ˈplæzmə]
n. 等离子体；血浆；

facilitated diffusion
[fəˈsɪləˈteitid dɪˈfjʊʒən]
n. 促进扩散

active transport
[ˈæktɪv ˈtrænspɔːt]
n. 主动运输

group translocation
[gruːp ˌtrænsləʊˈkeɪʃən]
n. 基团转位

energy, carbon, nitrogen, phosphorus, sulfur, and various minerals, the precise composition of a satisfactory medium will depend on the species one is trying to cultivate because nutritional requirements vary so greatly. Knowledge of a microorganism's normal habitat often is useful in selecting an appropriate culture medium because its nutrient requirements reflect its natural surroundings. Frequently a medium is used to select and grow specific microorganisms or to help identify a particular species. In such cases the function of the medium also will determine its composition.

参考译文

1. 一般的营养需求

对微生物细胞成分的分析表明：微生物细胞干重的95%以上由C、O、H、N、S、P、K、Ca、Mg和Fe少数几种主要元素组成，这些元素称为大量元素或大量营养物质，因为微生物对它们的需要量相对较大。C、O、H、N、S和P是糖类、脂类、蛋白质及核酸的主要成分，而Ca、Mg和Fe主要以离子形式存在于细胞中，它们的生理功能是多种多样的。例如，K^+是许多酶（包括蛋白质合成过程中涉及的一些酶）的必需组分；Ca^{2+}在细菌芽孢的耐热性方面起重要作用；Mg^{2+}可作为一些酶的辅助因子，可与ATP结合，并有助于核糖体及细胞膜的稳定；Fe^{2+}和Fe^{3+}不仅是细胞色素的组分，也可作为一些酶及电子载体蛋白辅助因子的组分。

除了大量元素外，所有的生物，包括微生物，还需要一些微量元素或微量营养物质。大部分细胞需要Mn、Zn、Co、Mo、Ni和Cu等微量元素，而且细胞所需的量很少。水、玻璃容器及一般的培养基组分中都含有微量元素，常常能满足生长的需求，因此要精确确定微生物对微量元素的营养需求十分困难。在自然界，微量营养物质无所不在，通常不会限制生长。微量元素一般作为酶或辅助因子的组分辅助催化反应，维护蛋白质结构的稳定。

除了大量元素和微量元素外，一些具有特殊形态结构和在特殊环境下生长的微生物具有特殊的营养需求。例如，硅藻需要硅酸来合成其富含二氧化硅的细胞壁。虽然大多数细菌不需要大量的钠，但一些生活在盐湖及海洋中的细菌依靠高浓度的钠离子（Na^+）生存。

最后必须强调的是：微生物需要一个平衡的混合营养，假如必需的营养物质供应不足，无论其他营养物质的浓度怎样，微生物的生长将受到限制。

2. 微生物的营养类型

生物生长除需要碳、氢和氧之外，还需要能源和电子。根据微生物对所有这些需求的不同，可将它们分为不同的营养类型。根据所利用碳源的不同，微生物可分为异养型或自养型。对生物而言，仅有两种能源可以利用：

（1）光能。

（2）分解有机物或无机物获取的化学能。

光能营养型微生物利用光能，而化能营养型微生物从化合物或有机物或无机物的氧化作用中获得能量（可分为无机营养和有机营养型）。硫细菌不能氧化水，而利用 H_2、H_2S 和 S 等无机物为电子供体。化能有机异养型微生物（通常称异养菌）所需的能源、电子供体（氢供体）和碳源都来自有机物。通常情况下，同一种有机物可满足所有这些需要。值得注意的是，所有致病微生物本质上都是化能有机异养型。

相比之下，光能有机异养型和化能无机自养型微生物较少，但是它们常常在生态学上非常重要。光能有机异养型微生物可以利用光能并以有机物作为电子供体和碳源，如某些紫绿细菌。光能有机异养型微生物通常生活在被污染的湖泊和河流中，其中的某些种类也可利用 H_2 作为电子供体而以光能无机自养型的方式生活。化能无机自养型微生物 CO_2 为碳源，通过氧化还原无机物，如铁、氮或硫分子，获得能源和电子用于生物合成。少数化能无机自养型微生物可以利用有机物获得碳源，以异养的方式生活。化能无机自养型微生物在生态系中的元素化学转化过程中不断起重要作用。例如，在氨转变为硝酸盐，以及硫转变成硫酸盐的过程中都有这类微生物的参与。

3. 细胞对营养物质的吸收

微生物细胞对营养物质的利用是从对它们的吸收开始的，这种吸收机制是专一性的，也就是说，只吸收需要的物质，吸收不能利用的物质对细胞不利。微生物通常生活在营养物质贫乏的环境中，因此它们必须具有将营养物质从胞外低浓度环境运输到胞内高浓度环境的逆浓度运输能力，而且细胞质膜必须具有选择透过性，允许营养物质进入胞内，而阻止其他物质自由进入。由于营养物质的多样性和复杂性，微生物有多种方式对营养物质进行运输。其中最重要的几种包括促进扩散、主动运输和基团转位。在真核微生物中尚未发现基团转位，它们可以通过胞吞作用吸收营养物质。

4. 培养基

微生物学的许多研究取决于在实验室培养和保存微生物的能力，而这只有有效地利用适宜的培养基才有可能。用于微生物生长、移植和保存的培养基可以是固体或液体制品，有效的培养基必须含有微生物生长所需的全部营养物。另外，一些特定的培养基可用于微生物分离、鉴定、测定抗生素效价、检测水质和食品质量及工业生产等。尽管所有微生物在具有能源、碳、氮、磷、硫和无机盐的条件下可以生长，但由于不同微生物具有不同的营养需求，培养基的具体组成要依微生物各自的营养需求特点而定。对微生物自然生长环境中营养条件的了解有助于在实验室配制合适的培养基，因为这种特定的自然环境营养条件反映了生活在该处的微生物的天然的营养需求。培养基常用于筛选、培养和鉴定某一特定微生物，常根据其功能来决定其组成。

Further Reading

The Development of Culture Media

The earliest culture media were liquid, which made the isolation of bacteria to prepare pure cultures extremely difficult. In practice, a mixture of bacteria was **diluted** successively until only one organism, as an average, was present in a culture vessel. If everything went well, the individual bacterium thus isolated would reproduce to give a pure culture. This approach was tedious, gave variable results, and was plagued by contamination problems. Progress in isolating pathogenic bacteria understandably was slow.

The development of techniques for growing microorganisms on solid media and efficiently obtaining pure cultures was due to the efforts of the German bacteriologist Robert Koch and his associates. In 1881 Koch pub-fished an article describing the use of boiled potatoes, sliced with a flame-sterilized knife, in culturing bacteria. The surface of a sterile slice of potato was inoculated with bacteria from a needle tip, and then the bacteria were streaked out over the surface so that a few individual cells would be separated from theremainder. The slices were incubated beneath bell jars to. Koch decided to try solidifying this medium. Koch was an amateur photographer-he was the first to take photomicrographs of bacteria and was experienced in preparing his own photographic plates from silver salts and **gelatin**. Precisely the same approach was employed for preparing solid media. He spread a mixture of Loeffler's medium and gelatin over a glass plate, allowed it to harden, and inoculated the surface in the same way he had inoculated his sliced potatoes.

The new solid medium worked well, but it could not be incubated at 37℃ (the best temperature for most human bacterial **pathogens**) because the gelatin would melt. Furthermore, some bacteria digested the gelatin.

About a year later, in 1882, agar was first used as a solidifying agent. It had been discovered by a Japanese innkeeper, Minora Tarazaemon. The story goes that he threw out extra seaweed soup and discovered the next day that it had jelled during the cold winter night. Agar had been used by the East Indies Dutch to make jellies and jams. Fannie Eilshemius Hesse, the New Jersey-born wife of Walther Hesse, one of Koch's assistants, had learned of agar from a Dutch acquaintance and suggested its use when she heard of the difficulties with gelatin. Agar-solidified medium was an instant success and continues to be essential in all areas of microbiology.

New Words

dilute [daɪˈluːt] *vt.* 稀释；冲淡；削弱 *adj.* 稀释的；淡的

gelatin [ˈdʒelətɪn] *n.* 明胶；动物胶；胶制品

pathogens [ˈpæθədʒəns] *n.* 病原体；病原菌；致病菌

参考文献

[1] Atlas. R. M. Handbook of microbiological media, 2d ed. Boca Raton, Fla.: CRC Press, 1997.

[2] Bridson. E. Y. Media in microbiology. Rev. Med. Microbial. 1990, 1:1–9.

[3] Cote. R. J., Gherna. R. L. Nutrition and media. In Methods for general and molecular bacteriology 2d ed. Washington, D.C.: American Society for Microbiology, 1994.

[4] Difco Laboratories. Dif'co manual dehydrated culture media and reagents for microbiology. 11th ed. Sparks, Md.: BD Bioscience, 1998.

[5] Power. D. A. Manual of BBL products and laboratory procedures, 6th ed. Cockeysville, Md.: Becton, Dickinson and Company, 1998.

Chapter 3　Microbial Growth　微生物的生长

Growth may be defined as an increase in cellular constituents. It leads to a rise in cell number when microorganisms reproduce by processes like budding or **binary fission**. In the latter, individual cells enlarge and divide to yield two **progeny** of approximately equal size. Growth also results when cells simply become longer or larger. If the microorganism is **coenocytic** that is, a **multinucleate** organism in which nuclear divisions are not accompanied by cell divisions growth results in an increase in cell size but not cell number. It is usually not convenient to investigate the growth and reproduction of individual microorganisms because of their small size. Therefore, when studying growth, microbiologists normally follow changes in the total population number.

1. The **growth curve**

Population growth is studied by analyzing the growth curve of a microbial culture. When microorganisms are cultivated in liquid medium, they usually are grown in a batch culture or closed system-that is, they are incubated in a closed **culture vessel** with a single batch of medium. Because no fresh medium is provided during incubation, nutrient concentrations decline and concentrations of wastes increase. The growth of microorganisms reproducing by binary fission can be plotted as the **logarithm** of the number of viable cells versus the incubation time. The four phase are **lag phase, exponential phase, stationary phase, death phase.**

New Words and Expressions

binary fission ['baɪnəri 'fɪʃən]
n. （原生动物、细胞等的）二分裂；二分体
progeny ['prɒdʒəni]
n. 子孙；后裔；成果
coenocytic [siːnə'sitik]
adj. 多核细胞的
multinucleate [ˌmʌlti'njuːkliit]
adj. 多核的

growth curve [grəuθ kəːv]
n. 生长曲线，增长曲线
culture vessel ['kʌltʃə 'vɛsl]
n. 培养瓶；培养皿
logarithm ['lɒgərɪðəm]
n. 对数
lag phase ['læg fez]
n. 延滞期
exponential phase ['ɛkspə'nɛnʃəl fez]
n. 对数生长期
stationary phase ['steʃənɛri fez] n. 稳定期
death phase [deθ feiz]
n. 衰亡期

2. The continuous culture of microorganisms

(1) solutes and **water activity** Most bacteria, **algae**, and fungi have rigid cell walls and are **hypertonic** to the habitat because of solutes such as amino acids, **polyols**, and potassium ions. The amount of water actually available to microorganisms is expressed in terms of the water activity (aw). Although most microorganisms will not grow well at water activities below 0.98 due to **plasmolysis** and associated effects, **osmotolerant** organisms survive and even flourish at low aw values. Halophiles actually require high sodium chloride concentrations for growth.

(2) Temperature Environmental temperature profoundly affects microorganisms, like all other organisms. Indeed, microorganisms are particularly susceptible because they are usually **unicellular** and their temperature varies with that of the external environment. For these reasons, microbial cell temperature directly reflects that of the cell's surroundings. A most important factor influencing the effect of temperature on growth is the temperature sensitivity of enzyme-catalyzed reactions. At low temperatures a temperature rise increases the growth rate because the velocity of an enzyme-catalyzed reaction, like that of any chemical reaction, will roughly double for every 0℃ rise in temperature. Because the rate of each reaction increases, metabolism as a whole is more active at higher temperatures, and the microorganism grows faster. Beyond a certain point further increases actually slow growth, and sufficiently high temperatures are **lethal**. High temperatures damage microorganisms by denaturing enzymes, transport carriers, and other proteins.

water activity [wɔːtə æek'tiviti]
n. 水分活度；水分活性
algae ['ældʒiː]
n. 藻类
hypertonic [ˌhaɪpə'tɒnɪk]
adj. 高渗的
polyols [pəʊl'jɒlz]
n. 多元醇
plasmolysis [plæz'mɒlɪsɪs]
n. 质壁分离
osmotolerant [ɒzməʊtəʊlərənt]
adj. 渗透耐性的

unicellular [ˌjuːnɪ'seljʊlə]
adj. 单细胞的

lethal ['liːθəl]
adj. 致命的，致死的

Microbial membranes are also disrupted by **temperature extremes**; the **lipid bilayer** simply melts and disintegrates. Thus, although functional enzymes operate more rapidly at higher temperatures, the microorganism may be damaged to such an extent that growth is **inhibited** because the damage can-not be repaired. At very low temperatures, membranes solidify and enzymes don't work rapidly. In summary, when organisms are above the **optimum temperature**, both function and cell structure are affected. If temperatures are very low, function is affected but not necessarily cell chemical composition and structure. The temperature dependence of enzyme activity.

Because of these opposing temperature influences, microbial growth has a fairly characteristic temperature dependence with distinct **cardinal** temperatures-minimum, optimum, and maximum growth temperatures. Although the shape of the temperature dependence curve can vary, the temperature optimum is always closer to the maximum than to the minimum. The cardinal temperatures for a particular species are not rigidly fixed but often depend to some extent on other environmental factors such as pH and the available nutrients.

The cardinal temperatures vary greatly between microorganisms. **Optima** normally range from 0℃ to as high as 75℃ whereas microbial growth occurs at temperatures extending from 20℃ to over 100℃ The major factor determining this growth range seems to be water. Even at the most extreme temperatures, microorganisms need liquid water to grow. The growth temperature range for a particular microorganism

temperature extreme
[ɪksˈtriːm ˈtempərɪtʃə]
n. 温度极限
lipid [ˈlɪpɪd]
n. 脂质；油脂
bilayer [ˈbaɪleɪə]
n. 双分子层（膜）
inhibited [ɪnˈhɪbɪtɪd]
v. 抑制；控制 adj. 抑制的
optimum temperature
[ˈɔptɪməm ˈtempərɪtʃə]
n. 最适温度；最佳温度

cardinal [ˈkɑːdɪnəl]
adj. 主要的，基本的；深红色的

optima [ˈɔptɪmə]
n. 最适条件；最佳状态

usually **spans** about 30 degrees. Some species (e.g., *Neisseria gonorrhoeae*) have a small range; others, like *Enterococcus* faecalis, will grow over a wide range of temperatures. The major microbial groups differ from one another regarding their maximum growth temperature. The upper limit for protozoa is around 50 ℃. Some algae and fungi can grow at temperatures as high as 55 to 60℃. Procaryotes have been found growing at or close to 100℃, the boiling point of water at sea level. Recently strains growing at even higher temperatures have been discovered. Clearly, procaryotic organisms can grow at much higher temperatures than eucaryotes. It has been suggested that eucaryotes are not able to manufacture **organellae** membranes that are stable and functional at temperatures above 60℃. The **photosynthetic apparatus** also appears to be relatively unstable because photosynthetic organisms are not found growing at very high temperatures.

(3) **Oxygen concentration** Microorganisms can be placed into at least five different categories based on their response to the presence of O_2: **obligate aerobes, facultative anaerobes,** aerotolerant anaerobes, strict or obligate anaerobes, and microaerophiles. Oxygen can become toxic because of the production of hydrogen peroxide, superoxide radical, and hydroxyl radical. These are destroyed by the enzymes **superoxide dismutase, catalase,** and **peroxidase**.

(4) **Radiation** High-energy or short-wavelength radiation harms organisms in several ways. **Ionizing radiation**-X rays and gamma rays-ionizes molecules and destroys DNA and other cell components. Ultraviolet (UV) radiation induces the formation of **thymine dimers** and strand breaks

spans [spænz]
v. 跨越；持续；贯穿
Neisseria gonorrhoeae [naiˈsiəriə ˌgɒnəˈrɪə]
n. 淋球菌；淋病奈瑟菌
Enterococcus [ˌentərəʊˈkɒkəs]
n. 肠球菌
organellae [ɔːgeɪˈneliː]
n. 细胞器蛋白
photosynthetic apparatus [ˌfəʊtəsɪnˈθetɪk ˌæpəˈrætəs]
n. 光合器官
oxygen concentration [ˈɒksɪdʒən ˌkɒnsənˈtreɪʃən]
n. 氧浓度
obligate aerobe [ˈɒblɪˌgeɪt ˈeərəʊb]
n. 专性需氧微生物
facultative anaerobe [ˈfækl̩ˌtetɪv æˈneɪərəʊb]
兼性厌氧菌
superoxide dismutase [ˌsupərˈɒksaɪd dɪsmuteɪs]
n. 超氧化物歧化酶
catalase [ˈkætəleɪz]
n. 过氧化氢酶
peroxidase [pəˈrɒksɪdeɪz]
n. 过氧物酶
ionizing radiation [ˈaɪənaɪzɪŋ ˌredɪˈeʃən]
n. 电离辐射
thymine dimer [ˈθaɪmɪn ˈdaɪmə]
n. 胸腺嘧啶二聚体

in DNA. Such damage can be repaired by photoreactivation or dark reactivation mechanisms. (5) pH Each species of microorganism has an optimum pH for growth and can be classified as an **acidophile**, **neutrophile**, or **basophil**. Microorganisms can alter the pH of their surroundings, and most culture media must be buffered to stabilize the pH.	acidophile [əˈsidəfail] n. 嗜酸细胞 neutrophile [ˈnjuːtrəfɪl] n. 嗜中性细胞 basophil [ˈbeɪsəʊfɪl] n. 嗜碱细胞

参考译文

生长就是指细胞组分的增加。对以出芽或二分分裂进行繁殖的微生物来说，生长就会导致细胞数量增加，细胞个体增长到一定程度就分裂成两个大小基本相等的子代细胞。对多核微生物而言，细胞核的分裂并不伴随细胞分裂，生长意味着细胞体积增加而个体数目不变。由于微生物个体微小，以个体为对象研究其生长和繁殖十分不便，常以群体数量来研究这类微生物的生长。

1. 生长曲线

对微生物群体生长的研究是通过分析微生物培养物的生长曲线来进行的。通常将微生物置于一个封闭系统中，比如在液体培养基中进行分批培养。该封闭系统实际上就是将培养基置于一个封闭的容器内，接入微生物，在培养过程中不更换培养基进行，随着培养时间延长，营养物质浓度下降，而新陈代谢产生的废物增加。若以培养时间为横轴，以细胞数量的对数为纵轴，就可做出一条反映以二分分裂方式进行繁殖的微生物生长规律的生长曲线，该曲线由4个阶段组成，分别是延滞期、对数期、稳定期、衰亡期。

2. 环境因素对微生物生长的影响

（1）溶质与水活度   大多数细菌、藻类和真菌具有坚韧的细胞壁，胞内具有氨基酸、多元醇及K^+等溶质，相对于周围环境而言，胞内为高渗环境，可被微生物利用的水量的多少用水活度 Aw 表示。由于质壁分离和其他影响，大多数微生物在水活度低于 0.98 的情况下生长不好，但耐渗微生物可在低水活度条件下生长繁殖，而嗜盐菌需要高 NaCl 浓度才能生长。

（2）温度   就像对其他生物一样，环境的温度对微生物也有很大影响。事实上，由于微生物通常是单细胞型生物，它们的温度随周围环境温度的变化而变化，所以它们对温度的变化特别敏感。正因为如此，微生物细胞温度也直接反映了所处环境的温度。温度对微生物生长影响的一个决定性因素是微生物酶催化反应对温度的敏感性。在低温条件范围内，温度升高可加快生长速度，因为酶催化反应与一般的化学反应一样，反应速度随温度每提高10℃而加倍，由于胞内各种反应都加速，整个新陈代谢活动在较高的温度下更加活跃，微生物生长更快。当温度升高到一定程度时，继续升温会使生长速度下降，而过高的温度会导致微生物死亡，因为在高温条件下，微生物酶、运输载体及其他蛋白会发生热变性，细胞脂质双分子层膜在高温下熔化崩解，从而使细胞受到损害。因此，尽管酶催化反应在高温下进行得更快，但由于上述原因使细胞受到难以恢复的损伤，导致生长受到抑制。在很低的温度下，细胞膜会冻结，酶也不能迅速工作。总之，如果超过了微生物生长的最适温度，其功能和细胞结构均会受影响；如果温度很低，虽然功能受到影响，但是细胞的化学组分和结构不一定全受影响。

由于温度对微生物生长有利与不利两方面的共同影响，微生物生长具有相当明显的温度依赖性，有最低、最适和最高生长温度这几个基本温度。尽管微生物不同或条件不同时生长的温度依赖曲线会有变化，但最适温度总是更靠近最高生长温度而不是最低生长温度。同一种微生物的三个基本温度并不是一成不变，而是在一定程度上依赖于 pH 及营养物质可获得性等其他因素。

不同微生物相互之间基本温度有很大差别，最适生长温度可低至 0℃ 或高达 75℃，能够进行生长的温度低至 −20℃，高的甚至超过 100℃。决定这种生长范围的主要因素似乎是水，甚至在最极端温度条件下微生物也需要液态水才能生长。对一种微生物而言，其生长温度范围幅度一般为 30℃ 以内，某些种类如淋病奈瑟菌，其生长温度范围很窄。而像粪肠菌等可在一个很宽的温度范围生长。微生物几个主要类群各自的最高生长温度相互之间有差别，原生动物最高生长温度为 50℃ 左右，一些藻类和真菌可在 55～60℃ 条件下生长。原核生物可在接近或等于沸点的条件下生长，近来还发现有的细菌能在高于 100℃ 条件下生存。显然，原核生物比真核生物更能在较高的温度条件下生长，据认为这是因为真核生物在高于 60℃ 的条件下不能构建稳定且具有相应功能的细胞器膜。此外，光合作用细胞器似乎也相当不稳定，因为还没有发现能在高温下生长的光合生物。

(3) 氧浓度　根据与氧之间关系的不同，可将微生物分为专性好氧菌、兼性厌氧菌、耐氧厌氧菌、专性厌氧菌和微好氧菌。氧对专性厌氧菌和微好氧菌的毒害作用是因为其不能产生超氧化物歧化酶和过氧化氢酶而降解过氧化氢、超氧化物自由基。

(4) 辐射　高能或短波辐射通过多种方式损害生物，X 射线和 Y 射线等电离辐射使分子发生电离，并破坏 DNA 和其他细胞组分。紫外辐射诱导胸腺嘧啶二聚体的形成，并导致 DNA 发生断裂，而光复活作用和暗复活作用可修复这种损伤。

(5) pH　每种微生物都有各自的生长最适 pH，因而有嗜酸菌、嗜中性菌和嗜碱菌之分。微生物能改变周围环境的 pH，大多数培养基都具有缓冲系统以维持 pH 的稳定。

Further Reading

Temperature for Growth

Until recently the highest reported temperature for procaryotic growth was 105℃. It seemed that the upper temperature limit for life was about 100℃ the boiling point of water. Now thermophilic procaryotes have been reported growing in sulfide chimneys or "black smokers", located along rifts and ridges on the ocean floor, that spew sulfide-rich **superheated** vent water with temperatures above 350C. Evidence has been presented that these microbes can grow and reproduce at 113℃. The pressure present in their habitat is sufficient to keep water liquid (at 265atm; seawater doesn't boil until 460℃).

The implications of this discovery are many. The proteins, membranes, and nucleic acids of these procaryotes are remarkably temperature stable and provide ideal subjects for studying the ways in which macromolecules and membranes are stabilized. In the future it may be possible to design enzymes that can operate at very high temperatures. Some **thermostable** enzymes from these organisms have important industrial and scientific uses. For example, the Taq **polymerise** from the thermophile. Thermos aquaticus is used extensively in the polymerise chain reaction

New Words:

superheated [ˌsupəˈhitɪd] *adj.* 过热的，过热蒸汽

thermostable [θɜːməʊˈsteɪbəl] *adj.* 热稳定的；耐热性的

polymerise [ˈpɒlɪməraɪz] *vt.* 使聚合（等于 polymerize）

参考文献

[1] Atlas. R. M. Handbook of microbiological media, 2d ed. Boca Raton, Fla.: CRC Press, 1997.

[2] Bridson. E. Y Media in microbiology. Rev. Med. Microbial. 1990, 1:1–9.

[3] Cote. R. J., Gherna. R. L. Nutrition and media. In Methods for general and molecular bacteriology 2d ed. Washington, D.C.: American Society for Microbiology, 1994.

[4] Difco Laboratories. Dif'co manual dehydrated culture media and reagents for microbiology. 11th ed. Sparks, Md.: BD Bioscience, 1998.

[5] Power. D. A. Manual of BBL products and laboratory procedures, 6th ed. Cockeysville, Md.: Becton, Dickinson and Company, 1988.

Chapter 4   Recent Developments in the Production of Valuable Microbial Products   重要微生物发酵产品的最新进展

As the scientist-writer Steven Jay Gould emphasized, we live in the Age of Bacteria. They were the first living organisms on our planet, live virtually everywhere life is possible, are more numerous than any other kind of organism, and probably constitute the largest component of the earth's **biomass**. The whole ecosystem depends on their activities, and they influence human society in countless ways. Thus modern microbiology is a large discipline with many different specialties; it has a great impact on fields such as medicine, agricultural and food sciences, ecology, **genetics**, biochemistry, and molecular biology.

For example, microbiology has been a major contributor to the rise of molecular biology, the branch of biology dealing with the physical and chemical aspects of living matter and its function. Microbiologists have been deeply involved in studies on the genetic code and the mechanisms of DNA, RNA, and protein synthesis. Microorganisms were used in many of the early studies on the **regulation** of gene expression and the control of enzyme activity. In the 1970s new discoveries in microbiology led to the development of **recombinant DNA technology** and genetic engineering.

One indication of the importance of microbiology in the twentieth century is the Nobel Prize given for work in physiology or medicine. About 1/3 of these have been awarded to scientists working on microbiological problems. Microbiology

New Words and Expressions

biomass ['baɪəʊmæs]
n.（单位面积或体积内的）生物量

genetics [dʒɪ'netɪks]
n. 遗传学

regulation [regjʊ'leɪʃən]
n. 管理；规则；校准

recombinant DNA technology
[ri'kambənənt DNA tɛk'nalədʒi]
n. DNA 重组技术

has both basic and applied aspects. Many microbiologists are interested primarily in the biology of the microorganisms themselves. They may focus on a specific group of microorganisms and be called **virologists** (viruses), bacteriologists (bacteria), phycologists or algologists (algae), **mycologists** (fungi), or protozoologists (protozoa). Others are interested in microbial morphology or particular functional processes and work in fields such as microbial **cytology**, microbial physiology, microbial ecology, microbial genetics and molecular biology, and microbial taxonomy. Of course a person can be thought of in both ways (e.g., as a bacteriologist who works on **taxonomic** problems). Many microbiologists have a more applied **orientation** and work on practical problems in fields such as medical microbiology, food and dairy microbiology, and public health microbiology (basic research is also conducted in these fields). Because the various fields of microbiology are interrelated, an applied microbiologist must be familiar with basic microbiology. For example, a medical microbiologist must have a good understanding of microbial taxonomy, genetics, immunology, and physiology to identify and properly respond to the pathogen of concern.

  What are some of the current occupations of professional microbiologists? One of the most active and important is medical microbiology, which deals with the diseases of humans and animals. Medical microbiologists identify the agent causing an infectious disease and plan measures to eliminate it. Frequently they are involved in tracking down new, unidentified pathogens such

virologist [vaɪə'rɔlədʒəst]
n. 病毒学家
mycologist [maɪ'kɔlədʒɪst]
n. 真菌学家；霉菌学家

cytology [saɪ'tɒlədʒɪ]
n. 细胞学

taxonomic [ˌtæksəu'nɔmik]
adj. 分类的；分类学的
orientation [ˌɔːrɪən'teɪʃən]
n. 方向；定向；适应

as the agent that causes variant **creutzfeldt-Jacob** disease, the **hanta virus**, and the virus responsible for AIDS. These microbiologists also study the ways in which microorganisms cause disease.

Public health microbiology is closely related to medical microbiology. Public health microbiologists try to control the spread of communicable diseases. They often monitor community food establishments and water supplies in an attempt to keep them safe and free from infectious disease agents.

Immunology is concerned with how the immune system protects the body from pathogens and the response of infectious agents. It is one of the fastest growing areas in science; for example, techniques for the production and use of **monoclonal antibodies** have developed extremely rapidly. Immunology also deals with practical health problems such as the nature and treatment of **allergies** and autoimmune diseases like **rheumatoid arthritis**.

Many important areas of microbiology do not deal directly with human health and disease but certainly contribute to human welfare. Agricultural microbiology is concerned with the impact of microorganisms on agriculture. Agricultural microbiologists try to combat plant diseases that attack important food crops, work on methods to increase soil **fertility** and crop yields, and study the role of microorganisms living in the **digestive tracts** of **ruminants** such as cattle. Currently there is great interest in using bacterial and viral insect pathogens as substitutes for chemical **pesticides**.

The field of microbial ecology is concerned with the relationships between microorganisms and their living and nonliving habitats. Microbial ecologists

---

creutzfeldt-Jacob *n.* 克雅病
hanta virus [ˌhæntə ˈvaiərəs]
n. 汉坦病毒

immunology [ɪmjʊˈnɒlədʒɪ]
n. 免疫学

monoclonal antibodies
[ˌmɔnəˈkləʊnəl ˈæntɪˌbɔdiː]
n. 单克隆抗体；单株抗体
allergy [ˈælədʒɪ]
n. 过敏症；反感；厌恶
rheumatoid arthritis
[ˈrʊmətɔɪd ɑrˈθratis]
n. 类风湿性关节炎

fertility [fəˈtɪlɪtɪ]
n. 多产；肥沃；生产力；丰饶
digestive tract
[daɪˈdʒestɪv trækt]
n. 消化道
ruminant [ˈruːmɪnənt]
n. 反刍动物
pesticides [ˈpestɪsaɪdz]
n. 农药；杀虫剂

study the contributions of micro-organisms to the carbon, nitrogen, and sulfur cycles in soil and in fresh water. The study of pollution effects on microorganisms also is important because of the impact these organisms have on the environment. Microbial ecologists are employing microorganisms in **bioremediation** to reduce pollution effects.

Scientists working in food and dairy microbiology try to prevent microbial spoilage of food and the transmission of food borne diseases such as **botulism** and salmonellosis. They also use microorganisms to make foods such as cheeses, yogurts, **pickles**, and beer. In the future microorganisms themselves may become a more important nutrient source for livestock and humans.

In industrial microbiology microorganisms are used to make products such as antibiotics, **vaccines**, steroids, alcohols and other solvents, vitamins, amino acids, and enzymes. Microorganisms can even leach valuable minerals from low-grade **ores**.

Research on the biology of microorganisms occupies the time of many microbiologists and also has practical applications. Those working in microbial physiology and biochemistry study the synthesis of antibiotics and toxins, microbial energy production, the ways in which microorganisms survive harsh environmental conditions, microbial nitrogen fixation, the effects of chemical and physical agents on microbial growth and survival, and many other topics.

Microbial genetics and molecular biology focus on the nature of genetic information and how it regulates the development and function of cells and organisms. The use of microorganisms has been

bioremediation ['baiəuri‚miːdi'eiʃən]
n. 生物修复；生物降解

botulism ['bɒtjʊlɪzəm]
n. 肉毒中毒（食物中毒一种）

pickle ['pɪkəl]
n. 泡菜；盐卤；腌制食品

vaccine ['væksiːn]
n. 疫苗；牛痘苗

ore [ɔː]
n. 矿；矿石

very helpful in understanding gene function. Microbial geneticists play an important role in applied microbiology by producing new microbial strains that are more efficient in synthesizing useful products. Genetic techniques are used to test substances for their ability to cause cancer. More recently the field of genetic engineering has arisen from work in microbial genetics and molecular biology and will contribute substantially to microbiology, biology as a whole, and medicine. Engineered microorganisms are used to make **hormones,** antibiotics, vaccines, and other products. New genes can be inserted into plants and animals; for example, it may be possible to give corn and wheat nitrogen fixation genes so they will not require **nitrogen fertilizers.**

hormone ['hɔːməun]
n. 激素；荷尔蒙；性激素；荷尔蒙制剂
nitrogen fertilizer ['naitrədʒən 'fəːtilaizə]
n. 氮肥；氮肥肥料

参考译文

科学家、作家 Steven Jay Gould 指出，我们生活在细菌的时代，它们是我们星球上最先出现的生命有机体。所有角落都有他们的身影，其数量比任何其他种类的有机体都多，可能是地球生物总量的最大组成部分。全部生态系统都依赖于它们的活动，它们影响人类社会的各个方面。所以现代微生物学是一个具有许多不同专业方向的大学科，它对医学、农学和食品科学、生态学、遗传学、生物化学和分子生物学都有重大影响。

微生物学为分子生物学的兴起做出了主要贡献。分子生物学是论述生命物质的物理和化学及其功能的生物学分支学科。微生物学家已深入地参与了遗传密码及 DNA、RNA 和蛋白质合成机理的研究。在早期基因表达的调节和酶活性的控制研究中，许多方面都是利用微生物进行的。在 20 世纪 70 年代，微生物学中新的发现导致 DNA 重组技术和遗传工程的发展。

20 世纪微生物学重要性的象征之一是在诺贝尔生理学或医学奖奖项获得者中约 1/3 是关于微生物学问题研究的科学家。微生物学既有基础理论方向又有应用方向，许多微生物学家主要感兴趣的是微生物本身的生物学。由于他们的研究焦点是微生物的特有类群而被划分为病毒学家、细菌学家、藻类学家、真菌学家或原生动物学家。另外一些微生物学家主要研究微生物形态学或特殊的功能过程，他们的研究领域包括微生物细胞学、微生物生理学、微生物生态学、微生物遗传学和分子生理学及微生物分类学。当然也有同时在这两个领域展开研究的

（如作为研究分类学问题的细菌学家）。许多微生物学家有较多的应用方向，他们在诸如医学微生物学，食品和乳品微生物学，公共卫生微生物学（在这些领域也有基础性的研究）等领域中从事实际问题的研究。因为微生物学的各种领域是相互联系的，所以一位应用微生物学家必须熟悉基础微生物学，例如，一位医学微生物学家为了鉴定和正确地了解相关的病原体，他必须很好地懂得微生物分类学、遗传学、免疫学和生理学知识。

职业微生物学工作者现在从事一些什么工作呢？最活跃和最重要的领域之一是医学微生物学，它与人和动物的疾病打交道，医学微生物学家鉴定引起感染疾病的因子，并设计措施将其消灭。他们要经常地跟踪新的未辨别出的病原体，例如，引起变异性克雅病的因子、汉坦病毒属、引起艾滋病的病毒。这些微生物学工作者也研究微生物引起疾病的途径，如军团杆菌病、汉坦病毒肺综合征和艾滋病。

公共卫生微生物学与医学微生物学是紧密相关的，公共卫生微生物学家尝试控制传染病的传播，他们经常检验食品和供应的水，努力保持食品和水的安全，并且不受传染疾病因子的影响。

免疫学涉及免疫系统如何预防病原体保护身体和应答感染因子，它是科学中发展最迅速的领域之一，例如，单克隆抗体产生技术及其应用已极其迅速地发展。免疫学也研究实际的健康问题，例如，研究通过单克隆抗体治疗过敏的本质和治疗像类风湿性关节炎一类的自身免疫病。

微生物学的许多重要领域并不直接研究人类的健康和疾病，但对人类的福利却做出了重大贡献。农业微生物学涉及微生物对农业的影响，农业微生物学家研究如何防止损害重要作物的植物疾病，采取措施提高土壤肥力提高庄稼产量，研究生活在牛一类反刍动物消化道中微生物的作用，当前，利用昆虫的细菌和病毒病原体作为化学农药的代用品，已引起人们很大的兴趣。

微生物生态领域涉及微生物与它们栖息地的生命体和无生命体之间的相互关系。微生物生态学家研究微生物对土壤和淡水中碳、氮和硫循环的贡献。关于微生物对污染作用的研究也是重要的，因为这些有机体已对环境产生了影响，微生物生态学家将微生物应用于生物整治中以减少污染的作用。

在食品和乳品微生物学方面，科学家致力于预防食物的腐败和食物传播传染病的研究，例如，肉毒中毒和沙门菌病；他们也利用微生物制造食品，例如，乳酪、酸牛乳、泡菜和啤酒。将来，微生物本身可能成为家畜和人类较重要的一种营养资源。

在工业微生物学中，用微生物来生产产品，例如，抗生素、疫苗、类固醇、醇和其他试剂、维生素、氨基酸和酶。微生物甚至能从低品质的矿物中筛选出有价值的矿物。

许多微生物学家致力于微生物的生物学研究和其在实际中的应用，例如，微生物生理和生物化学方面的工作主要是研究抗生素和毒素的合成、微生物能源的产生、微生物存活于苛刻环境中的方式、微生物固氮、化学和物理因子对微生物生长和存活的影响等课题。

微生物的遗传学和分子生物学的研究重点是遗传信息的本质及其如何调节细胞和有机体的发育和其他功能。微生物对于认识基因功能有很大的帮助。微生物遗传学家通过培育能更有效地合成有用产物的新菌株而在应用微生物学中起着重要作用。遗传学技术也被用来检测可能引起癌症的物质。遗传工程领域已经从微生物的遗传学和分子生物学的工作基础上兴起，并将对微生物学、整个生物学和医学做出很大贡献，基因工程微生物被用于生产激素、抗生素、疫苗和其他的产品，新基因能够插入植物和动物，例如，将固氮基因导入谷物和小麦，使它们不需要施氮肥。

Further Reading
The Future of Microbiology

What are some of the most promising areas for future microbiological research and their potential practical impacts? What kinds of challenges do microbiologists face? The following brief list should give some idea of what the future may hold:

(1) New infectious diseases are continually arising and old diseases are once again becoming widespread and destructive. AIDS, **hemorrhagic** fevers, and tuberculosis are excellent examples of new and reemerging infectious diseases. Microbiologists will have to respond to these threats, many of them presently unknown.

(2) Microbiologists must find ways to stop the spread of established infectious diseases. Increases in antibiotic resistance will be a continuing problem, particularly the spread of multiple drug resistance that can render a pathogen impervious to current medical treatment. Microbiologists have to create new drugs and find ways to slow or prevent the spread of drug resistance. New vaccines must be developed to protect against diseases such as AIDS. It will be necessary to use techniques in molecular biology and **recombinant** DNA technology to solve these problems.

(3) Research is needed on the association between infectious agents and chronic diseases such as **autoimmune** and cardiovascular diseases. It may be that some of these chronic afflictions partly result from infections.

(4) We are only now beginning to understand how pathogens interact with host cells and the ways in which diseases arise. There also is much to learn about how the host resists pathogen invasions.

(5) Microorganisms are increasingly important in industry and environmental control, and we must learn how to use them in a variety of new ways. For example,

microorganisms can (a) serve as sources of high-quality food and other practical products such as enzymes for industrial applications, (b) degrade pollutants and toxic wastes, and (c) be used as vectors to treat diseases and enhance agricultural productivity. There also is a continuing need to protect food and crops from microbial damage.

(6) Microbial diversity is another area requiring considerable research. Indeed, it is estimated that less than 1% of the earth's microbial population has been cultured. We must develop new isolation techniques and an adequate classification of microorganisms, one which includes those microbes that cannot be cultivated in the laboratory. Much work needs to be done on microorganisms living in extreme environments. The discovery of new microorganisms may well lead to further advances in industrial processes and enhanced environmental control.

(7) Microbial communities often live in biofilms, and the sebiofilms are of profound importance in both medicine and microbial ecology. Research on biofilms is in its infancy; it will be many years before we more fully understand their nature and are able to use our knowledge in practical ways. In general, microbe-microbe interactions have not yet been extensively explored.

(8) The genomes of many microorganisms already have been sequenced, and many more will be determined in the coming years. These sequences are ideal for learning how the genome is related to cell structure and what the minimum assortment of genes necessary for life is. Analysis of the genome and its activity will require continuing advances in the field of bioinformatics and the use of computers to investigate biological problems.

(9) Further research on unusual microorganisms and microbial ecology will lead to a better understanding of the interactions between microorganisms and the inanimate world. Among other things, this understanding should enable us to more effectively control pollution. Similarly, it has become clear that microorganisms are essential partners with higher organisms in symbiotic relationships. Greater knowledge of symbiotic relationships can help improve our appreciation of the living world. It also will lead to improvements in the health of plants, livestock, and humans.

(10) Because of their relative simplicity, microorganisms are excellent subjects for the study of a variety of fundamental questions in biology. For example, how do complex cellular structures develop and how do cells communicate with one another and respond to the environment?

(11) Finally, microbiologists will be challenged to carefully assess the implications of new discoveries and technological developments. They will need to

communicate a balanced view of both the positive and negative long-term impacts of these events on society.

The future of microbiology is bright. The microbiologist Rene Dubos has summarized well the excitement and promise of microbiology: How extraordinary that, all over the world, microbiologists are now involved in activities as different as the study of gene structure, the control of disease, and the industrial processes based on the phenomenal ability of microorganisms to decompose and synthesize complex organic molecules. Microbiology is one of the most rewarding of professions because it gives its practitioners the opportunity to be in contact with all the other natural sciences and thus to contribute in many different ways to the betterment of human life.

New Words

hemorrhagic ['hemərædʒɪk] *adj.* 出血的

recombinant [rɪ'kɒmbɪnənt] *n.* 重组；重组体

autoimmune [ˌɔːtəʊɪ'mjuːn] *adj.* 自身免疫的；自体免疫的

参考文献

[1] Atlas. R. M. Probiotics-snake oil for the new millennium Environ. Microbial. 2000, 1:375-382.

[2] Chang. S., Buswell. J. A., et al.Genetics and breeding of edible mushrooms. New York: Gordon and Breach, 1993.

[3] Fritts. C. A., Kersey. J. H., et al.Bacillus subtilis C-3102 (Calsporin) improves live performance and microbiology, 2000.

[4] Dixon. B. Microbiology present and future. ASM News. 1997, 63 (3):124-125.

[5] Young. P. American academy of microbiology outlines basic research priorities. ASM News. 1997, 63 (10):546-550.

Unit 4　Fermentation Engineering　发酵工程

Chapter 1　Fermentation Microbiology and Biotechnology　微生物发酵和生物技术

1. The nature of fermentation

The origins of fermentation technology were largely with the use of microorganisms for the production of foods and **beverages** such as cheeses, **yoghurts,** alcoholic beverages, vinegar, **sauerkraut**, fermented pickles and sausages, soya sauce and the products of many other Oriental fermentations (Table4-1). The present-day large scale production processes of these products are essentially scale-up versions of former domestic arts. Paralleling this development of product formation was the recognition of the role microorganisms could play in removing unpleasant wastes and this has resulted in massive world-wide service industries involved in water purification, effluent treatment and waste management. New dimensions in fermentation technology have made use of the ability of microorganisms. ① to overproduce specific essential primary metabolites such as **glycerol, acetic acid, lactic acid,** acetone, **butyl alcohol, butanediol,** organic acids, amino acids, vitamins, polysaccharides and xanthans; ② to produce useful secondary metabolites (groups of metabolites that do not seem to play an immediate recognizable role in the life of the microorganism producing them) such as **penicillin**, streptomycin, oxytetracycline, **cephalosporin**,

New Words and Expressions

beverages [be'vərɪdʒɪz]
n. 饮料

yoghurt ['jəʊgɜːt]
n. 酸乳酪，酸乳；酸牛乳

sauerkraut [saʊəkraʊt]
n. 酸菜；腌菜；酸泡菜

glycerol ['glɪsərɒl] n. 甘油

acetic acid [əˌsiːtɪk 'æsɪd]
n. 乙酸，醋酸

lactic acid [ˌlæktɪk 'æsɪd]
n. 乳酸

butyl alcohol ['bjuːtil 'ælkəˌhɒl]
n. 丁醇

butanediol [bjuːteɪ'niːdaɪəl]
n. 丁二醇

penicillin [penɪ'sɪlɪn]
n. 青霉素，盘尼西林

cephalosporin [sefələʊ'spɔːrɪn]
n. 头孢菌素

belligerents, alkaloids, actinomycin; and ③ to produce enzymes as the desired industrial product such as the exocellular enzymes amylases. proteases, pectinases or intracellular enzymes such as invertase, **asparaginase,** uric oxidase restriction endonucleases and DNA ligase. More recently, fermentation technology has begun to use cells derived from higher plants and animals under conditions known as cell or tissue culture. Plant cell culture is mainly directed towards secondary product formation such as alkaloids, perfumes and flavours while animal cell culture has initially been mainly concerned with the formation of protein molecules such as interferons, monoclonal antibodies and many others.

  Future markets are largely assured for fermentation products because, with limited exceptions, it is not possible to produce them economically by other chemical means. Furthermore, economies in production will also occur by genetically engineering organisms to unique or higher productivities. The commercial market for products of fermentation technology is almost united but will ultimately depend on economics and safety considerations.

  The processes of commercial fermentation are in essence very similar no matter what organism is selected, what medium is used and what product formed. In all cases, large numbers of cells with uniform characteristics are grown under defined, controlled conditions. The same apparatus with minor modifications can be used to produce an enzyme, an antibiotic, an organic chemical or a single cell protein. In its simplest form fermentation processes can be just the mixing of microorganisms with a

asparaginase
[æspə'rædʒineis]
n. 天（门）冬酰胺酶

nutrient broth and allowing the components to react. More advanced and sophisticated large-scale processes require control of the entire environment so that the fermentation process can proceed efficiently and, what is more important, can be exactly repeated with the same amounts of raw materials, broth and cell inoculum producing precisely the same amount of product (Table 4-1).

nutrient [njuːtriənt]
n. 营养物，营养品，养分，养料；滋养物

Table 4-1　Fermentation products according to industrial sectors

Sector	Activities
Chemicals Organic (bulk)	Ethanol, acetone, butanol
	Organic acids (citric, itaconic)
	Enzymes
Organic (fine)	Perfumeries
	Polymers (mainly polysaccharides)
Inorganic	Metal beneficiation, **bioaccumulation** and leaching
Pharmaceuticals	Antibiotics
	Diagnostic agents (enzymes, monoclonal antibodies)
	Enzyme inhibitors
	Steroids
	Vaccines
Energy	Ethanol (gasohol)
	Methane (biogas)
	Biomass
Food	Dairy products (cheeses, yogurts, fish and meat products)
	Beverages (alcoholic, tea and coffee)
	Baker's yeast
	Food additives (antioxidants, colours, flavours, stabilizers)
	Novel foods (soy sauce, **tempeh, miso**)
	Mushroom products
	Amino acids, vitamins
	Starch products
	Glucose and high fructose syrups
	Functional modifications of proteins, pectins

(continued)

bioaccumulation [baɪəʊəkjuːmjʊˈleɪʃən]
n. 生物体内积累

tempeh [tempə]
n. 印尼豆豉；霉大豆
miso [miːsəʊ]
n. 日本豆面酱

Sector	Activities
Agriculture	Animal feed stuffs (SCP) Veterinary **vaccines** Ensilage and composting processes Microbial pesticides Rhizobium and other N-fixing bacterial inoculants Mycorrhizal inoculants Plant cell and tissue culture (vegetative propagation, **embryo** production, genetic improvement)

vaccines [væk'siːnz]
n. 疫苗

embryo [embriəʊ]
n. 胚芽

All biotechnological processes are carried out within a containment system or bioreactor. The physical form of most common bioreactors has not altered much over the past 30 years. Recently, however, many novel forms have been designed and they may play an increasingly active part in biotechnology. The main function of a bioreactor is to minimize the cost of producting a product or service while the driving force behind the continued improvement in design and the quality of the product or service. Studies have considered better **aseptic** design and operation, better process control including computer involvement, and how to obtain a better understanding of the rate-limiting steps of a system, especially heat and mass transfer.

aseptic [ə'septɪk]
adj. 无菌的，经消毒的

In biotechnology, processes can be broadly considered to be either conversion cost intensive or recovery cost intensive. With conversion cost intensive processes the volumetric productivity, Qp (kg of product $m^{-3}h^{-1}$), is of major importance while with recovery cost intensive processes the product concentration, P(kgm^{-3}), is the main **criterion** for e minimization of cost. Examples of the diverse

criterion [krai'tiəriən]
n. （批评、判断等的）标准，准则，规范

product categories produced in bioreactors in the biochemical process industry are given in Table 4-2.

Table 4-2 Examples of products in different categories in biotechnological industries

Category	Example
Cell mass	Baker's yeast, single cell protein
Cell components	Intracellular proteins
Biosynthetic products	Antibiotics, vitamins, amino and organic acids
Catabolic products	Ethanol, methane, lactic acid
Bioconversion	High-fructose **corn syrup**, **6-aminopenicillanic acid**
Waste treatment	**Activated sludge**, anaerobic digestion

There are three main operating types of bioreactors for biotechnological processes together with two forms of biocatalysts. Bioreactors can be operated on a batch, semicontinuous (**fed-batch**) or continuous basis. Reactions can occur in static or agitated cultures, in the presence or absence of oxygen, and in aqueous or low moisture conditions (solid substrate fermentations). The biological catalysts can be free or can be attached to surfaces by immobilization or by natural **adherence**. The **biocatalysts** can be cells in a growing or non-growing state or isolated enzymes used as soluble or immobilized catalysts. In general, the reactions occurring in a bioreactor are conducted under moderate conditions of pH (near neutrality) and temperature (20 to 65℃). In most bioreactors the processes occur in an **aqueous phase** and product streams will be relatively dilute.

corn syrup [kɔːn 'sirəp]
n. 玉米糖浆

6-aminopenicillanic acid
n. 6-氨基青霉烷酸

activated sludge ['æktɪveɪtɪd slʌdʒ]
n. 活性污泥

fed-batch [fid bætʃ]
n. 补料分批

adherence [æd'hɪərəns]
n. 依附

biocatalyst [biˈəʊkətəlɪst]
n. 生物催化剂

aqueous phase ['eikwiəs feiz]
n. 水相

The optimization of a bioreactor process involves minimizing the use of raw materials (e.g. Nutrients, precursors, acid/base, air) and energy (the cost of energy since 1978 has risen at an annual rate of 16%), and maximizing product purity and quality in the broth before recovery. Process optimization is achieved by manipulation of both the physical and chemical parameters associated with the process. The range of process variables that are important to process development and are discussed.

2. Chemical and biological engineering

Biological engineering and bioengineering are no longer new topics for chemical engineering. However, while these terms have become quite common in the lexicon of the discipline, the concepts associated with them are still rather ill defined and often confusing. As there is no longer doubt about the growing importance of biology as enabling science of industries critical for chemical engineering, it is becoming imperative that a discussion be initiated to eventually lead to a consensus about the meaning of biological engineering and its relevance to the profession of chemical engineering. This consensus could then guide present activities in **curriculum** reform, department name changes, reorientation of student and faculty interests and similar efforts by our professional organization. The purpose of this perspective is to offer some thoughts that might be helpful to this end.

Let me begin by noting that the most precious asset of a profession is its intellectual core. In an era of rapid evolution of industry and curricula, it is imperative that a discipline defines its own core

curriculum [kəˈrikjuləm]
n. 全部课程，课程

and strengthens it through scholarly activity and diverse applications. During the past 15 years, we have witnessed an increasing attraction of students and faculty to biological applications of chemical engineering. This trend is paralleled by an on-going transformation of the chemical processing industry to a life sciences based one. In **response** to these developments, we have expanded our base to include biology, along with chemistry and physics, as fundamental science of chemical engineering. However, it is important to note that, amidst such drastic industrial and educational changes, our own chemical engineering core remains virtually unchanged regardless of its application.

response [ri'spɔns] *n.* 反应

It is important to note that the vision of chemical and biological engineering does not imply a radical departure from the present but an evolution into a very promising future. It does not change the core; on the contrary, it strengthens it with examples from a most exciting science. It does not suggest that areas where chemical engineers have contributed in the past and presently lead be abandoned. On the contrary, it enriches the portfolio of chemical engineering education and research, thus creating unparalleled opportunities at the interfaces between biology and the more traditional chemical engineering areas, such as, **polymers** (biomaterials), reaction engineering (metabolic engineering), computers and systems (systems biology), **thermodynamics** (thermo of large molecules), separations (of peptide fragments for proteomic applications), control (of metabolic pathways), (bio) catalysis, fluid mechanics (microfluidics for analytical applications), and many more.

polymers ['pɒlɪməs] *n.* 聚合物

thermodynamics [θɜːməʊdaɪ'næmɪks] *n.* 热力学

Unit 4 Fermentation Engineering 发酵工程

What is the best way to develop biological engineering?

The present national scene suggests that two models may be followed. In the first, chemical and biological engineering, and bioengineering (as defined earlier) are pursued in a parallel and mutually beneficial manner. The latter is centered around medical devices, imaging, biomechanics and biophysics, prostheses and radiology. The former encompasses medical and industrial applications of biological systems where chemical reactions are a fundamental determinant. It comprises multi-scale biological systems and applications ranging from the molecular to the cellular and equipment levels, such as, biotechnology, bioprocessing, metabolic engineering and cellular-molecular biomedical applications defined by equilibrium or rate processes. True to our attention to integration and quantification, I would also include that aspect of **bioinformatics** that deals with data integration aiming at elucidation of cellular function and physiology.

The second model consists of planting seeds from different curricula (including chemical engineering) into new organizational units to encourage the evolution of new versions of biological engineering free from the shackles of history. This approach is in direct opposition to all lessons learned from history about evolution of engineering disciplines in concert with industries. Nevertheless, it has a certain appeal with administrations eager to consolidate fragmented bioengineering efforts into a single department without regard to the need for an intellectual core shared by all faculty or an industry to absorb its graduates.

bioinformatics
[biːəʊɪnˈfɔːmətɪks]
n. 生物信息学

Until recently, chemical engineering departments were rather slow in embracing and moving forward with biological engineering. With few exceptions, they were content with a model of just a few bio-faculty within a steady-state, process-oriented, petro-centric paradigm. As it has been amply articulated before, this model is no longer satisfactory. In light of its ever-increasing importance, biological engineering cannot be viewed as another one of the many fields of chemical engineering. Biology should be included as a foundational science of our discipline, along with physics and chemistry. Curriculum reform is needed to reflect this fundamental change primarily in a **contextual** sense aiming at introducing chemical engineers to basic concepts of genetics, biochemistry and molecular cell biology. Such a curriculum reform would enrich chemical engineering as an engineering discipline and profession and will serve well the needs of the new chemical, biotechnology and **pharmaceutical** industries. We must work hard to infuse these concepts in our education and research and leave no doubt that the chemical engineering paradigm is the best vehicle for teaching and studying biological systems at the cellular and molecular levels.

Let there be no doubt that, should chemical engineering fail to respond to this challenge, there will be no lack of suitors to fill in the void. There is hardly any time left either. Although newly formed bioengineering departments are presently focusing on the more classical side of this field, it will not be too long before they take an interest in the cellular and molecular applications of biology

contextual [kən'tekstʃuəl]
adj. 文脉上的，前后关系的

pharmaceutical [fɑːmə'suːtɪkl]
adj. 制药的，配药的

and biotechnology. This trend will be implemented most likely by chemical engineers, who will not be developing a new version of bioengineering but will be simply transplanting the highly successful chemical engineering **paradigm** to bioengineering. I would suggest that such a development would be equivalent to robbing chemical engineering of its most precious resource, its intellectual core, with **devastating consequences** for the profession. There has been talk already of possible fragmentation of chemical engineering. A possible outcome of fragmentation may indeed be the demise of chemical engineering as we know it.

A fragmentation chemical engineering is not a legacy that will make us proud. We need to initiate without delay a serious curriculum development process to truly fuse biological content with the chemical engineering core and evolve our discipline into Chemical and Biological Engineering.

paradigm ['pærədaɪm]
n. 范例，样式，模范

devastating ['devəsteitiŋ]
adj. 毁灭性的，灾难性的，惊人的
consequences ['kɔnsikwənsiz]
n. 重要结果，重要地位，因果关系

参考译文

1. 发酵的本质

发酵技术的起源来自于大量利用微生物生产食品和饮料，像干酪、酸乳酪、酒精饮料、醋、泡菜、腌菜及腊肠、酱油和许多其他传统发酵食品。今天这些产品的大规模生产过程是过去家庭内部生产活动的放大版本。与产品形成的发展齐头并进的是对微生物作用的承认，这使得大规模世界范围服务业的出现，包括水的净化、污水处理及垃圾处理。发酵技术新的扩展来自于微生物的利用：① 过量生产重要的特殊的初级代谢产物比如甘油、醋酸、乳酸、丙酮、丁醇、丁二醇、有机酸、氨基酸、维生素、多糖和黄原胶；② 生产有用的次级代谢产物（代谢物群体其在生产它们的微生物的生命中发挥的作用好像不能很快地被认识到）像青霉素、链霉素、土孢菌素、头孢菌素、赤霉素、生物碱、放线菌素；③ 生产酶作为想要的工业产品，胞外酶像淀粉酶、蛋白酶、果胶酶或者胞内酶像转化酶、天冬酰胺酶、尿酸氧化酶、限制性核酸内切酶和DNA连接酶。最近，发酵技术开始利用高等植物和动物细胞进行我们所知道的细胞或组织培养。

植物细胞培养主要针对生产次级代谢物如生物碱、香料和调味品，而动物组织培养开始关注的是蛋白质分子形成如干扰素、单克隆抗体和许多其他的蛋白质。

这大大肯定了发酵产品的未来市场，由于生产有限，通过化学方法不能经济地生产这些产品。因此基因工程有机体是经济的而具有独特的和更高的生产能力。发酵技术产品的商业市场是无限的，但是最终要取决于经济与安全性方面的考虑。

发酵过程本质上非常相似的，不管选择的是什么有机体、用的是什么培养基及形成什么产物。在所有的情况下，大量的具有一致特征的细胞在限制的控制的条件下生长。同一个装置经过微小改动就可以用来生产酶、抗生素、有机化学试剂或者单细胞蛋白。发酵过程最简单的形式就是仅仅是营养微生物的混合，并使组分发生反应。更为先进和复杂的大规模生产过程需要对整体环境的控制从而使发酵过程能够有效地进行，更为重要的是，能够准确地进行重复，用相同量的原料、细胞组织生产出相同量的产物（表4-1）。

表4-1　　　　　　　　　　　　　工业发酵产品

类型	产品
化学品	乙醇，丙酮，丁醇
有机（散装）	有机酸（柠檬酸、衣康酸）
有机（整装）	酶
	香料
	聚合物（主要是多糖）
无机	金属选矿、生物浸出
中西药品	抗生素
	诊断试剂（酶、单克隆抗体）
	酶抑制剂
	类固醇
	疫苗
能量	乙醇（乙醇汽油）
	甲烷（沼气）
	生物量
食品	乳制品（干酪、酸乳、鱼、肉制品）
	饮料（含酒精、茶和咖啡）
	面包酵母
	食品添加剂（抗氧化剂、颜色、风味、稳定剂）
	新型食品（酱油、豆豉、豆酱）
	食用菌产品
	氨基酸，维生素
	淀粉制品
	糖和高果糖糖浆
	蛋白质的功能修改，果胶

续表

类型	产品
农业	动物性饲料（SCP） 兽用疫苗 青贮和堆肥工艺 微生物农药 根瘤菌固氮菌剂等 菌根菌剂 植物细胞和组织培养（植物繁殖、胚胎生产、遗传改良）

所有的生物工程过程都是在一个容器或者生物反应器中进行的。在过去的三十年里，大部分共同的生物反应器的物理形式没有发生多大的改动。然而，近来设计出了许多新型的生物反应器，它们将越来越积极地参与到生物工程中。生物反应器的主要功能是使产品或服务的生产成本最小化，同时保持设计的持续改进和产品或服务的质量背后的驱动力。开始研究考虑更好的设计和操作、更好的过程控制包括计算机的使用及怎样去更好地理解一个系统尤其是热量和质量转移系统的速度控制步骤。

在生物工程中，处理过程可认为是成本转化或者成本回收。对于转换成本，体积生产力 Q_p 是重要的，而对于转换成本，产品的浓度 P 是减少成本的主要标准。表4-2列出了生物化工工业利用生物反应器生产出的各种不同的产品。

表4-2　　　　　　　　生物技术产业的不同类别的产品实例

类别	举例
细胞团	面包酵母，单细胞蛋白
细胞成分	胞内蛋白
生物合成产物	抗生素、维生素、氨基酸和有机酸
分解代谢产物	乙醇、甲烷、乳酸
生物转化	高果糖玉米糖浆、6-氨基青霉烷酸
废物处理	活性污泥，厌氧消化

用于生物工程的生物反应器有三种主要的操作方式和两种生物催化形式。生物反应器可在分批式、半连续（分批补料）或者连续方式上进行操作。反应可以在稳定的或者搅动的培养液中，在有氧或者无氧、水溶液或者低湿度（固体底物发酵）条件下进行。可以是处于生长状态或者不处于生长状态的细胞或者是分离的酶用作游离的或固定的载体上。总体上，生物反应器中发生的反应是在温和的pH（近中性）和温度（20～65℃）条件下进行的。在大部分生物反应器里，反应过程是在水相中进行的，产品就相对被稀释了。

对生物反应器过程的优化包括减少原料（如养分、前体、酸/碱、空气）和能量（自1978年能量消耗以平均每年16%的速度上涨）的使用，在回收前提高产物的纯度和质量。过程优化是通过控制过程的物理和化学参数来实现的。

2. 化学与生物工程

对化学工程来说，与生物相关的工程和生物工程已不是新话题，一些术语在这个学科的词典里已变得相当普通，但相关概念的定义还不完全准确，往往较含糊。毫无疑问，对于化学工程相关的产业来说，生物学作为支撑学科的关键作用在日益增长，目前当务之急是发起一场讨论，从讨论中对生物工程的含义及其与化学工程这一关系上逐渐达成共识，并基于这个共识，指导课程的改革、专业和系名称的改变、学生和教授兴趣的重新定向，以及在学术协会组织方面的相关工作。本文旨在提供一些可能有助于上述目标的想法。

首先指出，对一个专业而言，其最宝贵的是它的知识体系的核心。在一个产业和学科课程都飞速发展的时代，一门学科必须定义自己的核心，并通过学术研究和各种不同的应用来强化这个核心。在过去15年里，我们看到学生和教员越来越被吸引到化学工程在生物方面的应用。与这一潮流平行的是，化学过程工业正转变成以生命科学为基础的过程工业。相应地，我们扩大了化学工程的基础，把生物学与化学和物理一起作为化学工程的基础学科。然而必须指出，在这些工业界和教育界的巨变中，化学工程的知识核心没有变，只是应用领域变了。

有必要指出，化学生物工程的内容和前景并不是激进地脱离现实，而是向着充满希望的前途发展。它并没有改变化学工程的知识核心。相反，一个个非常振奋人心的、科学的应用例子说明是强化了这一核心。新学科的建立并不意味着化学工程师们过去为之做出巨大贡献而现在仍领导着的领域将被抛弃；相反，它丰富了化学工程教育和科研的知识内容，并在生物学和传统化学工程领域的交叉面上创造出空前的发展机会，如聚合物（生物材料）、反应工程（代谢工程）、计算机和系统（系统生物学）、热力学（大分子热力学）、分离（蛋白质组学的多肽片段分离）、控制（代谢途径控制）、催化（生物催化）、流体力学（分析应用中的微流）等。

发展生物工程的最佳途径是什么？

目前国内（美国国内）的现状表明有两个模式可走。第一种模式以化学、生物工程与生物工程（如早期的定义）按平行和互利的方式发展。后者围绕着以医学设备、成像、生物力学和生物物理学、假体和放射医学方面的研究为中心发展。前者围绕着以化学反应为基本要素的生物体系在医学和工业上的应用；包括从分子水平、细胞水平到设备水平的多层次的生物知识系统及其应用，例如，生物技术、生物过程工程、代谢工程、由平衡或速率（反应和/或传递）过程决定的分子-细胞水平的生物医学应用。为了与我们对整合和量化的关注相一致,针对描述细胞功能和生理的研究,我还将把生物信息学中与数据整合方面的工作也包括在内。

第二种模式是从不同学科体系(包括化学工程)播撒种子,生长成新的不同知识单元,彻底摆脱历史的桎梏,促进全新的生物工程学科的发展。这个途径与历史上工程学科和工业协调发展的经验完全背道而驰。尽管如此,总还有人和某些机构热衷于把零散的生物工程整合成一个独立院系,而不顾及教员和吸收相关毕业生的产业部门对一个完整的知识核心的需求。

直到最近,化学工程部门在生物工程的衔接和推进中相当缓慢。几乎毫不例外地,他们乐意看到这样的模式:寥寥几个生物工程教授在一个稳定的、定位于过程的、以石油为中心的模式里运转。正如此前已清晰阐述的那样,这一模式不再适用。根据其空前的增长的重要性,生物工程不能仅仅被视为化学工程的众多分支之一。生物学应当作为化学工程学科的一门基础科学,就像物理和化学一样。必须进行课程改革来反映这一重大观念的改变,向化学工程师介绍遗传学、生物化学、分子细胞生物学的基本概念。这一课程改革将丰富化学工程的内容,作为学科和职业的化学工程,将更好地服务于新的化学、生物技术和制药产业。我们必须努力把这些概念融合到我们的教学和科研中去,并且让人们坚信,化学工程知识和新的模式是在细胞和分子水平上生物学知识体系进行教学和研究的最好载体。

毫无疑问,如果化学工程不能应对这一挑战,别的学科就会填补这一空白。所剩时间不多。尽管新组建的生物工程院系目前主要着眼于这一领域更加经典的方面,但不用多久,他们的兴趣就会转移到生物学和生物技术在细胞和分子水平上的应用研究上来。这种趋势很可能被化学工程师实现,他们并不是发展出一个新版的生物工程,而是简单地把非常成功的化学工程模式移植到生物工程领域。应指出的是,这种发展就像是把化学工程中最宝贵的财富——其知识核心抢走了,这对化学工程专业有破坏性的后果。

人们已经在谈论化学工程可能出现的分裂。分裂很可能造成的结果是化学工程的衰亡。一个分裂的化学工程难道真是我们想要的历史遗产吗?我们必须刻不容缓地发起一次严肃的课题改革,做到真正地把生物学知识融入到化学工程核心,把我们的学科发展成"化学生物工程"。

Further Reading

Fermentation in the China Beer Industry

This article examines the fast-evolving competitive dynamics that are now redefining the brewing industry in one of the world's largest markets——China. By examining the strategic approaches of **domestic** and multinational firms jockeying for market position in China, managers can better understand and act on some of the emerging critical success factors that surround market entry into a fast-developing country. As the Chinese economy continues its meteoric rise toward becoming another economic super-power, both domestic and multinational companies are discovering that competing in China is a double-edged sword. On the one hand, there is the promise of more than a billion consumers who are increasingly wealthy and crave new products. According to some estimates, tens of millions of Chinese enter the middle class each year. This means an enormous amount of purchasing power that well-positioned domestic and multinational firms hope to capture. On the other hand, the economic, regulatory, and even the **geographical** landscape of the country place an enormous burden on multinational firms seeking to build up their operations and supporting infrastructure. Although there are many cities in China that are beginning to display ultra-modern characteristics, vast portions of the country are left behind because of uneven economic development. The lack of transportation and communication systems complicates operational issues ranging from the supply-chain management to local labor pools.

The brewing industry in China crystallizes many of the paradoxes of competing in this enormous nation. Large, global players such as Anheuser-Busch Co., Heineken, SABMiller, and Interbrew have begun to invest heavily in the Chinese market in the past several years. These companies see China as the next frontier of rising per-capita beer consumption, which has leveled off or even declined in more traditional, mature markets, such as the U.S., Europe, and Canada. Yet, for many of these companies, entry into the Chinese market has proven difficult. Learning how to navigate the macroeconomic labyrinth of intricate distribution channels, poor transportation infrastructures, complex government regulations, and the seemingly endless number of domestic competitors of all sizes in every province and city has been an onerous task. All these companies have discovered that they must adapt their product and market strategies to accommodate the economic reality that there are indeed "many different Chinas" when serving the mainland "Chinese" market.

New Words

domestic [dəˈmestɪk] *adj*. 国内的；家庭的，家的；驯养的；热心家务的

geographical [dʒiːəˈgræfɪkl] *adj*. 地理学的，地理的

参考文献

[1] Saha B C. Lignocellulose Biodegradation and Applications in Biotechnology. Acs Symposium, 2004:2-34.

[2] El-Mansi E M T, Ward F B. Microbiology of industrial fermentation. In

Fermentation Microbiology and Biotechnology: CRC Taylor and Francis, London, New York (ISBN 0-8493-5334-3). //Book. 2007.

[3] El-Mansi E M T, Ward. F. B, et al. Microbiology of industrial fermentation. In Fermentation Microbiology and Biotechnology: CRC Taylor and Francis, London, New York ISBN 978-1-4398-5579-9. //BOOK. 2012.

Chapter 2　Strain Screen　菌种选育

1. Selection and screening

　　In all aspects of biotechnology major efforts are directed to screening programmes to genenrate new organisms, either from some natural source or from established cultures by way of mutation or by **hybridization** programmes, including genetic engineering. These organisms must be screened for useful products and grown up on a large enough scale to produce and extract the desired product and then to subject the product to critical evaluation. Screening can be defined as the use of highly selective procedures to allow the detection and isolation of only those **microorganisms** or metabolites of interest from a large population. The further **advancement** of an isolate will involve improvement and preservation of the culture. However, a major hindrance to the full exploitation of this capacity is the availability of suitable screeening procedures which can identify the necessary product, especially in the presence of culture medium constituents.

　　The major group of organisms presently used in biotechnonlogy are microoranisms. The screening **methodologies** to be described will concentrate largely on this group.

　　In the search for new microorganisms from the environment for biotechnological processes there are normally three types of option available; these involve the choice of habitat for sampling, the physical separation procedures for separating out the desired microorganism and the choice of method to achieve selection which in most cases involved **enrichment** cultures.

New Words and Expressions

hybridization [haɪbrɪdaɪˈzeɪʃn]
n. 杂化

microorganism [maikrəuˈɔːgənɪzəm]
n. 微生物

advancement [ədˈvɑːnsmənt]
n. 前进，进步；提升

methodology [mɛθədˈalədʒi]
n. 方法论

enrichment [ɪnˈrɪtʃmənt]
n. 丰富；改进，肥沃，发财致富

Unit 4 Fermentation Engineering 发酵工程

Although many new producer microorganisms are wildtypes and have been isolated from natural environments, major efforts are also directed to generating new genomes from existing genomes by laboratory manipulation. Organisms can be modified by **mutation, recombination, transformation, transduction** and gene cloning either by single processes or in combinations.

With natural selection and, more particularly, with genetic **manipulation**, al industrially important microorganisms will have been subjected to some form of screening. The design of the screening programme is of major importance to achieve maximum **recognition** of new genotypes. Screens can be divided into two basic forms. ① non-selective random screens with all the isolates being tested individually for the desired qualities; and ② rational screens in which there is some aspect of preselection.

Random screening can be very time-consuming because each isolate must be studied in detail. In studies of antibiotic production thousands of shake-flask cultures of single spore isolates are regularly used. Agar disc techniques have accelerated the process but this method is best used as a primary screen before using the shake flasks. Large numbers of isolates or mutants can now be plated out, exposed to a wide range of environmental conditions and the responses recorded automatically over any time period using television monitoring and computer control.

After screening procedures have been employed and potentially worthwhile microorganisms isolated the final proof of efficiency will come under production conditions. Product optimization from new and exising microorganisms also involves

mutation [mjuːˈteɪʃn]
n. 突变
recombination [riːˈkɒmbəˈneɪʃən]
n. 重组
transformation [trænsfəˈmeɪʃn]
n. 转化
transduction [trænsˈdʌkʃə]
n. 转导
manipulation [məˌnɪpjʊˈleɪʃ(ə)n]
n. 操纵；操作；处理；篡改
recognition [rekəgˈnɪʃən]
n. 识别

109

selection of the best method of cultivation, solid or liquid, batch or continuous. The type of medium used to **propagate** the organism can have a major influence on phenotypic expression of product formation. Thus medium composition can influence biomass concentration, specific rate of product formation, duration of product formation, duration of product formation, rate of decomposition and stability of the roducer organism.

Strain degeneration is a regular problem encountered in biotechnology. Stability is genetically controlled and can be modified by specific control of environmental factors such as medium or by the inhibition of the genetic events that lead to instability.

2. **Mutagenesis**

Once a useful organism has been isolated by preliminary screening several options are available for improving productivity, which include modifications to the medium and fermentation conditions, mutation (**spontaneous** or induced) and hybridization. Since mutaion is the ultimate source of all genetic variation it has become a topic of fundamental importance in industrial microbiology. Furthermore, many of the important microorganisms used industrially do not exhibit normal sexuality and therefore do not lend themselves easily to hybridization. For this reason there has been considerable use of various mutagenesis techniques to produce many of the currently used production strains. Indeed, for many industrial processes, mutagenesis programmes represent the main approach to increased productivity. Mutagen-induced improvements in productivity or titre have long been practised for

propagate ['prɒpəgeɪt]
vi. 繁殖；增殖

mutagenesis [ˌmjuːtə'dʒenɪsɪs]
n. 诱变

spontaneous [spɒn'teɪniəs]
adj. 自发的

antibiotic production in *Penicillin*, ***Cephalosporium*** and ***Streptomyces***, for organic acid production and enzyme production with Aspergillus species, amino acid production by various bacterial strains and for numerous other industrial microorganisms. Mutagenesis cannot yet be routinely supplanted by the newer techniques of genetic engineering.

It is now recognized that the most effective method for increasing the yield of a fermentation product is by use of induced mutations followed by selection of the improved strains. The main difficulty is that mutations occur at low frequencies and have to be selected from large non-mutant populations.

There are many examples of major mutational changes leading to improved industrial strains producting more efficacious products. Tetracycline producing strains of Streptomyces have been particularly significant in this area. A mutant strain of *S. aureofaciens* S604 was found to synthesize 6-demethylterracycline which was not synthesized by the parent production strain. This mutant is now one of the main commercial producers.

However, minor mutations have normally been the main approach to strain improvement. Such mutations normally lead to small improvements in product formation without any **phenotypic** manifestations. Successive use of small mutational changes can lead to an increase in the concentration of the hereditary factors responsible for the productivity of the original genotype. The **cephalosporin**-producing organisms have been of special interest here together with the penicillin producing *Penicillium chrysogenum* strains.

Cephalosporium [sefələʊˈspɔːrɪəm]
n. 顶头孢霉菌

Streptomyces [streptəʊˈmaɪsiːz]
n. 链霉菌

phenotypic [ˈfiːnətɪpɪk]
adj. 表型的

cephalosporin [sefələʊˈspɔːrɪn]
n. 头孢霉素

参考译文

1. 选择与筛选

在生物技术的所有方面中，最主要的精力都用在了筛选程序上，以通过突变或者杂交程序包括基因工程，从天然资源或者已经建立的培养物中生产出新的有机体。这些有机体必须经过筛选来获得有用的产品并且可以进行足够大规模的生产和提取所期望的产品，然后对这个产品进行鉴定评估。筛选可以定义为利用高效选择程序从庞大的微生物与代谢物中只检测和分离那些有益的微生物与代谢产物。对分离物进一步的改进包括对培养物的改造与保存。然而，对充分利用这个能力来说最大的障碍是适宜的筛选程序的可行性，这个筛选程序要求能鉴别所必需的产物，尤其在存在培养基组分的情况下。

现在生物工程中利用的主要的生物群体是微生物。所述的筛选方法也主要集中于这类群体。

在从环境中为生物工程寻找新的微生物的过程中，一般有三种可行性的选择；它们包括取样点的选择，分离出想要的微生物的物理分离过程以及实现选择的方法的选用，这个过程大多数情况下要进行富集培养。

尽管许多新产生的微生物是野生型而且已经从自然环境中分离出来，但是通过实验操作从现存的基因组制造新的基因组同样花费了大部分的精力。通过突变、重组、转化、转导和基因克隆的单一处理过程或者联合处理过程可以对有机体进行改造。

通过自然选择，尤其更多的是基因操作，所有工业重要的微生物都将经过某些形式的筛选。筛选程序的设计对实现最大程度的认知新基因性来说是很重要的。筛选可分为两个基本形式：① 无选择的随机筛选对所有分离物进行单独检测，以获得所要的性质 ② 有选择的筛选在这个过程中，会进行某些方面的预筛选。

随机筛选将是非常耗时的，因为每个分离物都要进行仔细的研究。在抗生素生产的研究中，成千上万的摇瓶培养的单一孢子分离物进行有规律的使用。琼脂板技术加快了这个过程，但这个方法最好是用作在使用摇瓶之前作为一种最初的筛选。大量的分离物或者突变体现在就可以平铺开来，暴露于广泛的环境条件下，同时任意时间段的反应就可以通过电视监控或者计算机控制而被自动记录下来。

当筛选过程完成获得有潜在价值的微生物分离后，最终的生产效率由生产条件表现。从新的或者现存的微生物进行产品的优化也涉及最佳培养条件的选择，固体或者液体、分批或者连续发酵。繁殖微生物的培养基类型对于形成产物的表型表达有着重要的影响。因此，培养基的组成能够影响生物体的浓度、形成产物的特定速率、产物形成的持续时间、产物的降解速率进行改造。

菌种的退化是生物工程中普遍遇到的一个问题。稳定性是由基因控制的，可以通过环境因素的特定控制进行改变，比如改变培养基或者通过阻止导致不稳定的遗传的发生。

2. 诱变

一旦一个有用的有机体经过最初的筛选被分离出来，那么对提高生产力的一些选择条件就可以使用了，包括对培养基和发酵条件的改变，诱变（自然发生或者是诱导发生）和杂交。既然诱变是所有遗传变化的根本来源，那么它就成为了在工业微生物学中有着根本重要性的一个话题。而且，这些工业中使用的很多重要的微生物没有正常的性特征，因此不易进行杂交。基于这个原因，各种各样的诱变技术就被大量采用以生产许多目前在使用的生产菌株。对于很多工业化过程，诱变程序确实代表着提高生产力所主要采用的方法。诱变剂——诱导提高生产力或者滴度的方法长期以来都在使用，青霉素、头孢霉素和链霉菌的抗生素生产，有机酸生产与用霉菌生产酶，用各种菌种生产氨基酸以及其他大量的工业微生物。诱变作用还无法被更新的基因工程技术所代替。

现在已经认识到，提高一个发酵产物产量最有效的方法是利用诱导突变和其后对改变菌株的筛选。最困难的是突变发生的频率低而且还必须从庞大的没有突变的菌株中进行选择。

有许多关于经重要突变所生产的改造工业菌株生产出更有效验的产品的例子。四环素的生产菌株链霉菌在这方面尤为重要。发现 S.aureofaciens 的突变株 S-604 可以合成去甲氯四环素脱甲基四环素，而不是由亲代生产菌株合成的。这个突变现在是一个重要的商业化程序。

尽管如此，很少的突变能够成为改造菌株的主要方法。这种突变通常对形成产物带来小的改变而没有任何表型表现。连续的利用小的突变可能导致负责原始基因型生产力的遗传因素的增加。生产头孢菌素的有机体与生产青霉素的菌株一样有着特殊的价值。

Further Reading

Screening Technologies

Ion channels are membrane spanning proteins which facilitates the rapid movement of inorganic ions across the cell membrane down their electrochemical gradient. Ion channels play a vital physiological role in neuronal signal transduction, neurotransmitter release, muscle contraction, cell secretion, enzyme activation, signal transduction and gene transcription. To date, upwards of 300 different human ion channel genes have been identified. Mutations in genes encoding ion channels can lead to channelopathies which result from a dysfunction of the altered ion channel protein. Thus, ion channels represent an important class of molecular targets for drug development.

A recent analysis of all FDA approved drugs available in 2002 has revealed that only 273 human genes have been used as the therapeutic targets. Ion channels correspond to 7% of the total recognized drug targets. It has been estimated that approximately 10% of the predicted 30,000 genes in human genome are druggable. Ion channel subunits and splice variants may represent 10 to 20% of the 3,000 predicted druggable gene targets.

The role of ion channels in drug safety has emerged as a critical issue in last several years. Since 1985, five drugs have been withdrawn from market due to their adverse effect on prolongation of the cardiac QT interval and in server cases, "torsade de pointes." The mechanisms underlying this toxic effect involve inhibition of one or more of the cardiac ion channels: ① human ether-go-go related gene (hERG), a potassium channel (Ikr); ② KCNQ1/CNE1 potassium channel (Iks); and SCN5A sodium channel.

Ion channels can be grouped into two classes: voltage-gated and **ligand** gated ion channels. Ion channels can exist in multiple states such as the closed, open and inactivated states. Voltage gated ion channels transition (gate) between these states in response to changes in membrane potential. Ligand-gated channels transition in between these states in response to the binding and unbinding of a ligand. In the open state, ions can flow through a single ion channel pore at prodigious rates of over 10^7 ions per second. Cell-based functional **assays** are an essential requirement for the screening of ion channels at both the primary and secondary levels. Traditional methods developed for high throughput screening of ion channels, such as binding, ion flux and fluorescent probes, measure ion channel activity indirectly. Patch clamp electrophysiology is regarded as the gold standard for measuring ion channel activity and pharmacology. Patchclamp allows for the direct, real time measurement of ion channel activity but in its traditional format is low throughput and requires a high degree of operator skill. Hence, drug screening assays for ion channels, in comparison with those for enzyme and receptor targets, have compromised data quality for throughput. However, a number of new screening technologies have been developed and improved for ion channel assays that are poised to change this.

New Words

Ion channel ['aɪən 'tʃænl] *n.* 离子通道

ligand ['lɪgənd] *n.* 配体；配合基

assay [ə'seɪ] *n.* 试验；化验

参考文献

[1] Turner M J, Courtney S M, et al. Strain Screen Suggests Genetic Influence on Physical Activity Varies Across the Lifespan. 2015, 47:79-80.

[2] Zhu S, Cai S, et al. Strain screen and condition optimization of synthesis arbutin by fermentation. China Brewing, 2012, 10 (6):46-49.

[3] Lightfoot J T, Leamy L, et al. Strain screen and haplotype association mapping of wheel running in inbred mouse strains. Journal of Applied Physiology, 2010, 109(3):623-634.

Chapter 3　Fermentation Medium　发酵培养基

1. Fermentation medium

The basic nutritional requirements of heterotrophic, microorganisms are an energy or carbon source, an available nitrogen source, inorganic **elements**, and water, for some microorganisms, growth factors and oxygen are also needed. The **elemental** composition of most microorganisms is fairly similar and consequently can be used as a starting point to designing an optimally balanced medium for fermentation. A variety of factors must be considered when formulating a fermentation medium. One relates to **cellular** stoichiometry and the desired amount of biomass to be produced. The basic concept here is simply a material balance; during the course of celluar growth, small organic and inorganic molecules such as glucose and ammonia are converted into biomass, Nutrients must be provided in sufficient quantities and proper proportions for a specified amount of biomass to be synthesized. **Computation** of these necessary amounts of various substrates clearly requires knowledge of the product composition.

Typical values for C. H. O. N and S as a percentage of dry cell weight are 45, 7, 33, 10, and 2.5, respectively.Trace elements such as Cu, Mn, Co, Mo, B and possibly other metal may be required, but the trace metals are normally present in sufficient amounts in the tap water or in the main raw materials. Some microorganisms require growth factors for their growth. Specific growth-factor supplements may be added to media as pure chemicals, for example,

New Words and Expressions

elements ['elɪmənts]
n. 成分；原理，基础

elemental [ˌelɪ'mentl]
adj. 基本的，主要的；自然力的

cellular ['seljələr]
adj. 细胞的；蜂窝状；由细胞组成的；多孔的

computation [ˌkɒmpju'teɪʃn]
n. 计算，估计

use of biotin in **glutamic** acid fermentations; or the growth-factors may be present as components of crude nitrogen sources such as corn steep liquor.

2. Growth medium

The microbes used for fermentation grow in (or on) specially designed growth medium which supplies the nutrients required by the organisms. A variety of media exist, but invariably contain a carbon source, a nitrogen source, water, salts, and micronutrients. In the production of wine, the medium is grape must. In the production of bio-ethanol, the medium may consist mostly of whatever inexpensive carbon source is available.

Carbon sources are typically sugars or other carbohydrates, although in the case of substrate transformations (such as the production of **vinegar**) the carbon source may be an alcohol or something else altogether. For large scale fermentations, such as those used for the production of ethanol, inexpensive sources of carbohydrates, such as **molasses**, corn steep liquor, sugar cane juice, or sugar beet juice are used to minimize costs. More sensitive fermentations may instead use purified glucose, sucrose, glycerol or other sugars, which reduces variation and helps ensure the purity of the final product. Organisms meant to produce enzymes such as beta **galactosidase**, **invertase** or other amylases may be fed starch to select for organisms that express the enzymes in large quantity.

Fixed nitrogen sources are required for most organisms to synthesize proteins, nucleic acids and other cellular components. Depending on the enzyme capabilities of the organism, nitrogen

glutamic [gluːˈtæmɪk]
adj. 谷氨酸的

vinegar [ˈvɪnɪɡər]
n. 醋

molasses [məˈlæsɪz]
n. 糖蜜；糖浆；废蜜

galactosidase [ɡəlæktəʊˈsaɪdeɪs]
n. 牛乳糖
invertase [ɪnˈvɜːteɪs]
n. 转化酵素；蔗糖转换酶

may be provided as bulk protein, such as soy meal; as pre-digested polypeptides, such as peptone or **tryptone;** or as ammonia or nitrate salts. Cost is also an important factor in the choice of a nitrogen source. Phosphorus is needed for production of **phospholipids** in cellular membranes and for the production of nucleic acids. The amount of phosphate which must be added depends upon the composition of the broth and the needs of the organism, as well as the objective of the fermentation. For instance, some cultures will not produce secondary metabolites in the presence of phosphate.

  Growth factors and trace nutrients are included in the fermentation broth for organisms incapable of producing all of the vitamins they require. Yeast extract is a common source of micronutrients and vitamins for fermentation media. Inorganic nutrients, including trace elements such as iron, zinc, copper, manganese, molybdenum and cobalt are typically present in unrefined carbon and nitrogen sources, but may have to be added when purified carbon and nitrogen sources are used. Fermentations which produce large amounts of gas (or which require the addition of gas) will tend to form a layer of foam, since fermentation broth typically contains a variety of foam-reinforcing proteins, peptides or starches. To prevent this foam from occurring or accumulating, **antifoaming** agents may be added. Mineral buffering salts, such as carbonates and phosphates, may be used to stabilize pH near optimum. When metal ions are present in high concentrations, use of a **chelating** agent may be necessary.

---

tryptone ['trɪptəʊn]
n. 胰化（蛋白）胨；胰胨

phospholipids [fɒspˈhɒlɪpɪdz]
n. 磷脂（类）

antifoaming [æntɪˈfəʊmɪŋ]
n. 消泡，防泡

chelating
v. 螯化的，有螯的，螯合的
(chelate 的现在分词)

参考译文

1. 发酵培养基

异养微生物需要能量或碳源,可用的氮源、无机元素和水,对于某些微生物、生长因子和氧也都是必要的。大多数微生物的元素组成是非常相似的,因此可以作为一个基础,设计一个最佳的培养基进行发酵。在制定发酵培养基时,必须考虑各种因素。一个涉及细胞的化学计量和所需的生物量所需的量。这里的基本概念是简单的物质平衡;细胞在生长的过程中,小的有机和无机分子,如葡萄糖和氨转化为生物量,营养必须充足、比例适当提供一定量的生物合成。计算这些必须的各种基质的量需要具备产品分子组成的知识。

主要元素包括 C、H、O、N,还有 S,这些元素占细胞干重的百分比分别是 45、7、33、10 和 2.5。元素如 Cu、Mn、Co、Mo、B 和其他可能的金属可能是必需的,但微量元素通常在自来水或主要原材料中存在足够量满足微生物要求。它们的特异性生长因子补充剂为纯化学品,例如,利用谷氨酸发酵生产时的生物素或可作为粗氮源的玉米浆作为生长因子。

2. 生长培养基

用于发酵的微生物在特定设计的生长培养基中生长(或在其上),其提供生物体所需的营养物。存在多种培养基,但是不变地培养基中都含有碳源、氮源、水、无机盐和微量元素。在葡萄酒的生产中,培养基是葡萄汁。在生物乙醇的生产中,培养基可以主要由任何便宜的碳源组成。

碳源通常是糖或其他碳水化合物,尽管在底物转化(例如生产醋)的情况下,碳源可以是醇或完全是其他物质。对于大规模发酵,例如用于生产乙醇的培养基,使用廉价的碳水化合物源如糖蜜、玉米浆、甘蔗汁或甜菜汁以使成本最小化。更敏感的发酵可以改为使用纯化的葡萄糖、蔗糖、甘油或其他糖,减少变化并帮助确保最终产物的纯度。要想产生酶如 β- 半乳糖苷酶、转化酶或其他淀粉酶的生物体可以进料淀粉以选择大量表达酶的生物体。

大多数生物体需要固定氮源来合成蛋白质,核酸和其他细胞组分。根据生物体的酶合成能力,氮源可以由散装蛋白如大豆粉提供;也可以由多肽消化分解提供,例如蛋白胨或胰蛋白胨;或作为氨或硝酸盐。成本也是选择氮源的重要因素。磷在细胞膜中产生磷脂和产生核酸是需要的。必须加入的磷酸盐的量取决于肉汤的组成和生物体的需要以及发酵的目的。例如,一些培养物在磷酸盐存在下不会产生次级代谢物。

生长因子和微量营养物包括在不能生产它们所需的所有维生素的生物体的发酵液中。酵母提取物是用于发酵培养基的微量营养素和维生素的常见来源。无机营养物,包括微量元素如铁、锌、铜、锰、钼和钴通常存在于未精制的碳和氮源中,但是当使用纯化的碳和氮源时可能必须添加。产生大量气体(或需要添加气体）的发酵将倾向于形成泡沫层，因为发酵肉汤通常包含多种蛋白，肽或淀粉使得泡沫增加。为了防止这种泡沫的发生或积聚，可以加入消泡剂。矿物缓冲盐，例如碳酸盐和磷酸盐，可用于稳定pH接近最佳。当金属离子以高浓度存在时，可能需要使用螯合剂。

Further Reading
Strategies for Improving Fermentation Medium Performance

When developing an industrial fermentation, designing a fermentation medium is of critical importance because medium composition can significantly affect product concentration, yield and volumetric productivity. For commodity products, medium cost can substantially affect overall process economics. Medium composition can also affect the ease and cost of downstream product separation, for example in the separation of protein products from a medium containing protein. There are many challenges associated with medium design. Designing the medium is a laborious, expensive, open-ended, often time-consuming process involving many experiments. In industry, it often needs to be conducted frequently because new mutants and strains are continuously being introduced. Many constraints operate during the design process, and industrial scale must be kept in mind when designing the medium. In Michael Crichton's fiction book, The Andromeda Strain, "The Wildfire project employed almost every known growth medium", totaling 80 in all. If only this were true! A medium design campaign can involve testing hundreds of different media. One of the more difficult aspects of the medium design process is dealing with this flow of data. In reality, often the information generated from design experiments is difficult to **assess** because of its sheer volume. Beyond about 20 experiments with five variables it is very difficult for a researcher to maintain medium **component** trends mentally, especially when more than one variable changes at a time. Data capture and data mining techniques are crucial in this situation.

Most of the studies published about medium improvement start with the objective of "given these components of the medium what is the best combination possible?" This can be referred to as a "closed strategy" in that it defines a fixed number of components and the type of components used. This is the simplest situation. The **disadvantage** of this strategy is that many different possible components, which are not considered, could be beneficial in the medium. It considers an extremely limited subset of design possibilities. It assumes you have chosen the right components to start with. The obverse situation, the "open strategy" asks, "What is the best combination of all possible components available?" This situation is much more complex and difficult to deal with. Experimental design strategies do not handle this situation easily. The advantage of the open strategy is that it makes no assumptions of which components are best. The ideal design strategy would be to start with an open strategy, and then move to a closed strategy once the best components have been selected. Too often researchers progress too quickly to a closed strategy. Three issues are particularly important to consider before medium design starts: the effect of medium design on strain selection; how well will shake flask medium design data scale up; and what is the target variable for improvement.

New Words

performance [pə'fɔːməns] n. 表现；表演；演技；执行

assess [ə'ses] vt. 评定；估价

component [kəm'pəʊnənt] n. 成分；组分；零件

disadvantage [dɪsəd'vɑːntɪdʒ] n. 不利，劣势，短处；损失

参考文献

[1] Kennedy M, Krouse D. Strategies for improving fermentation medium performance: a review. Journal of Industrial Microbiology & Biotechnology, 1999, 23(6):456-475.

[2] Rd F J, Davis B R. H7 antiserum-sorbitol fermentation medium: a single tube screening medium for detecting *Escherichia coli* O157:H7 associated with hemorrhagic colitis. Journal of Clinical Microbiology, 1985, 22(4):620-625.

[3] Cockshott A R, Hartman B E. Improving the fermentation medium for Echinocandin B production part II: Particle swarm optimization. Process Biochemistry, 2001, 36(7):661-669.

[4] Cockshott A R, Sullivan G R. Improving the fermentation medium for Echinocandin B production. Part I: sequential statistical experimental design. Process Biochemistry, 2001, 36(7):647-660.

Chapter 4 Fermenters 发酵罐

1. Fermenter

The fermenter is constructed of a suitable grade of **stainless** steel, with butt-welded joints polished flat on the inner surfaces. Over-lapping joints are unacceptable since the resulting ledge causes an accumulation of spent culture, creating problems on subsequent cleaning and **sterilization**: the high concentration of residual microorganisms requires longer heat-treatment to sterilize.

Most fermenters are fitted with stirring gear, which is particularly necessary to break up clumps of **mycelial** organisms. Various designs of agitator have been used, but the commonest is a set of flat vertical paddles set in a horizontal disc to create a circular movement of the medium. The **tendency** for vortex formation is prevented by vertical baffles fitted to the vessel walls, the bearing of the stirrer shaft has to support the mechanical stress of stirring a large volume of Viscous culture, but yet prevent access of microorganisms which might be drawn into the vessel by the rotation of the shaft. In the larger sizes of vessel, a continuously steam-sterilized section of the bearing prevents such **contamination**, with the additional benefit of lubrication by sterile condensate. The fermenter is steam sterilized (121℃, 15psi, 1h) with or without medium, which in the latter case is subsequently added. The sterile fermenter is filled to operational volume with sterile medium and a suitable cell **suspension** inoculum (1%~5% of the initial total culture volume). The inoculum is a stationary phase shake flask grown suspension

New Words and Expressions

stainless ['steɪnlɪs]
adj. 不锈的；不会脏的；无污点的；未被玷污的

sterilization [ˌsterəlaɪ'zeɪʃn]
n. 灭菌；杀菌，绝育

mycelial [maɪ'siːlɪəl]
adj. 菌丝体的

tendency ['tendənsi]
n. 倾向，趋势；旨趣，意向

contamination [kənˌtæmɪ'neɪʃn]
n. 污染；弄脏；毒害；玷污

suspension [sə'spenʃn]
n. 悬浮；悬架；悬浮液

culture. Sterile aeration and mixing rates were adjusted to the desired levels. The cultures were maintained at desired temperatures.

Actively metabolizing cultures generate heat: **attemperation** of fermentation vessels is essential. Smaller wessels, up to 1000L, can be conreolled effectively by a cooling jacket. With larger vessels the jacket/fermenter wall surface may be insufficient for effective cooling, and an internal cooling coil is necessary. However, a cooling coil, by providing horizontal surface within the vessel, is an additional complication to cleaning procedures. In modern equipment, cleaning is normally performed automatically by **spray** jets but it may be impossible to locate jets to clean all surfaces of a complex cooling coil system, and manual cleaning is the necessary.

2. Other fermenter design and ancillary equipment

A modified aerated STR (Stirred Tank Reactor), is used in vinegar production. In this design, air is drawn in and distributed via a high-speed hollow-body **turbin rotor**, connected to an air suction pipe. The aerator is self-**aspirating** and so some pressed air is not required.

The STR is the most widely used bioreactor for aerated fermentations perhaps because of its reliability and **flexibility.** Nevertheless, operating and investment costs are relatively high. In addition, problems exist in designing impellers for very large fermenters due to the length of the impeller shaft. In bioreactors without mechanical agitation, such as tower and loop bioreactors, aeration and mixing is achieved with high gas through-puts.

---

attemperation [ətem'pəreɪʃn]
n. 温度控制，调温

spray [spreɪ]
vt. 喷；喷射

turbin rotor ['tə:bin'rəʊtə]
n. 涡轮转子

aspirating ['æspərɪtɪŋ]
n. 除尘

flexibility [ˌfleksə'bɪlətɪ]
n. 柔度；柔韧性

In tower fermenters, air is introduced at the base of the fermenter and mixing is due to the rising bubbles so that the shear on the organism is minimal. Tower fermenters have been used to produce citric acid, by pellets of A.niger and by strains of Candida guillier mondii, and also for production of vinegar, industrial alcohol, and beer. Because vertical mixing is relatively poor in tower fermenters, they can be operated as continuous systems, with bottom entry feed and top exit, with high biomass **retention.** In non-aerated tower fermenters, used for brewing or industrial alcohol production, biomass retention is maximized by use of fermenting yeasts having good **flocculating** properties.

The loop air-lift bioreactor contains either an internal or external draft tube, sometime baffled, which increases mixing by forcing a directional flow of the bulk liquid. The driving force for circulation is created by the difference in density. Between the riser and down flow sections. In the ICI pressure cycle reactor, for example, air is introduced at the base of the fermenter and forced into solution by the **hydrostatic** pressure of the column. It will be seen that designs such as these have been used successfully for production of biomass by unicellular organisms. For cultivation of viscous filamentous mycelial cultures for SCP production combined draft tube impeller systems have been developed. In industrial fermentations involving animal and plant cells, which are more susceptible to shear damage than microbial cells, advantage has been taken of the lower shear effect of air-lift fermenters.

(1) Bench top or portable **autoclaves** Bench-top or portable autoclaves are useful for sterilizing

retention [rɪ'tenʃn]
n. 保留；记忆力

flocculating [f'lɒkjʊleɪtɪŋ]
v. 絮凝，絮结

hydrostatic [ˌhaɪdrə'stætɪk]
adj. 静水力学的，流体静力学的

autoclave ['ɔ:təʊkleɪv]
n. 高压釜；高压灭菌器；高压锅

small items, i.e. bottles, a small number of 500 mL conical flasks, automatic pipette tips, etc. Such autoclaves work on the same principle as a household pressure cooker. Programmable portable autoclaves are available but these are about six times as expensive as the standard portable autoclaves. The standard portable unit usually requires a form of external heating, although electrically heated models are available, and a gas burner should be purchased to supply that heat.

(2) **Incubators**, shakers Incubators are required for the cultivation of stock cultures. They range in size from controlled-temperature rooms to floor-standing units of varying size. Orbital incubators and shakers are required for the production of inocula, where a spore inoculum is not being used, and also for shake-flask experiment.

incubator ['ɪŋkjubeɪtər]
n. 孵化器，恒温箱

(3) **Ovens** Hot air ovens can be used for drying glassware or dry weight, or as sterilizing ovens. The working **chamber** should be made of corrosion-resistant stainless steel. Most hot air ovens rely on good air circulation; air is drawn into the oven, passes over a heating element and then passes out.

oven ['ʌvn]
n. 烤箱
chamber ['tʃeɪmbər]
n. 室

(4) **Pumps** At laboratory scale, liquid pumping is achieved almost exclusively by means of permanent pumps; diaphragm dosing pumps also find some application but are less flexible.

pumps ['pʌmps]
n. 泵

(5) **Membrane** type filters These have filter holding containing a cellulose acetate, or mitrate, or PTFE membrane of known and consistent pore size, which therefore retains particles larger than that pore size. 0.2μm or 0.45μm pore size filters are suitable for most application.

membrane ['membreɪn]
n. 膜

Packed-bed type filter (depth filters) have no uniform pore size and the **mechanism** of particle removal tends to be rather more complex. Typically, a filter housing is packed with glass wool or non-absorbent cotton wool. Such filters are vulnerable to compaction, and to wetting which may be allow channeling to occur. Sudden fluctuation in the pressure drop across the filter can cause release of particles or packing material under some circumstances.	mechanism ['mekənɪzəm] n. 机制，机能

参考译文

1. 发酵罐

发酵罐由一定规格的不锈钢对焊而成，对焊接头处的内表面需抛光处理。不能采用叠焊，因为叠焊的接头处产生缝隙，易积累发酵废液，对发酵罐的清洗和灭菌都不利，缝隙处残余的微生物越多，灭菌所需的时间就越长。

大多数发酵罐装配有搅拌装置，该搅拌装置对于打碎丝状菌所形成的菌球是非常必要的。所用的搅拌器的种类很多，最常见的类型是安装在水平圆盘上的平板垂直叶片式的，能使培养基产生循环运动。在容器壁上垂直安装挡板可防止涡流的产生。搅拌轴的轴承必须能承受搅拌大容量黏性培养液时的机械应力，同时能防止外界微生物趁轴在旋转时的机会吸入罐内。在大容量发酵罐中，在轴承上安装一套连续蒸汽灭菌装置可以防止上述污染，而且无菌的冷凝水对轴承还有润滑作用，可谓一举两得。发酵罐可在装有培养基的情况下，用蒸汽在121℃下灭菌 1 h，也可空罐灭菌。如果是这种情况，无菌培养基则在空罐灭菌后加到其工作容积，并接入适量的细胞悬浮种子液（初始培养液总体积的 1% ~ 5%）。菌种为处于平衡期的摇瓶培养液。无菌空气和搅拌速度应调节在适当的水平，发酵液维持在适当的温度。

代谢旺盛的培养液会产生大量的热，故有必要调节发酵罐的温度。1000L 以内的小型发酵罐用冷却夹套调温就够了。而对于大型发酵罐，夹套与发酵罐的壁表面太小，不能有效地冷却，则有必要安装罐内冷却蛇管。但是安装冷却蛇管后，蛇管的水平表面积增加，增加了清洗的难度。在现代发酵装置中，通常安装喷嘴进行自动清洗，但不管如何，所安装的喷嘴位置不可能清洗到结构极为复杂的冷却蛇管系统的所有表面，故人工清洗还是必不可少的。

2. 其他发酵罐设计和辅助设备

在醋生产中使用改进的充气搅拌罐反应器。在该设计中，空气通过连接到空气吸入管的高速空心体涡轮转子被吸入和分配。曝气器是自吸的，因此不需要一些压缩空气。

STR 是用于充气发酵中使用最广泛的生物反应器，或许是因为其可靠性和灵活性。然而，经营和投资成本相对较高。此外，由于叶轮轴的长度，在设计用于非常大的发酵罐的叶轮中存在问题。在没有机械搅拌的生物反应器中，例如，塔式循环生物反应器中，通过高气体通量实现通气和混合。

在塔式发酵罐中，在发酵罐的底部引入空气，并且由于上升的气泡混合，使得生物体上的剪切最小。塔式发酵罐已经用于通过黑曲霉和假丝酵母菌株产生柠檬酸，并且还用于生产醋，工业酒精和啤酒。因为塔式发酵罐中的垂直混合相对较差，所以它们可以作为连续系统操作，具有底部进料和顶部出料，具有高的生物量。在用于酿造或工业酒精生产的非充气塔式发酵罐中，通过使用具有良好絮凝性质的发酵酵母来最大化生物量保留。

环气升式生物反应器包含内部或外部引流管，有时会被阻挡，其通过强制主体液体的定向流动而增加混合。循环的驱动力由密度差产生。在提升管和下降流段之间。在 ICI 压力循环反应器中，例如，空气在发酵罐的底部引入，并通过塔的流体静压力而被迫进入溶液。可以看出，诸如这些的设计已经成功地用于通过单细胞生物体生产产品。为了培养用于 SCP 生产的黏性丝状菌丝体培养物，已经开发了合并的引流管叶轮系统。在涉及动物和植物细胞的工业发酵中，其比微生物细胞更易于受到剪切损伤，已经对空气提升发酵罐的较低剪切效应具有优势。

（1）台式或手提式压力蒸汽灭菌器　台式或便携式高压蒸汽灭菌锅可用于对小物品进行灭菌，如瓶、少量的 500mL 锥形瓶、自动移液枪头等。这种高压蒸汽灭菌锅与家用压力锅具有相同的原理工作。可编程便携式高压灭菌器是可用的，但这些是标准便携式高压灭菌器的六倍。标准的便携式设备通常需要一种外部加热的形式，尽管电加热型号是可用的，并且应当购买气体燃烧器以供应该热量。

（2）孵化器，振荡器　培养储存培养物需要孵育器。它们的尺寸范围从受控温度到不同大小的单位。需要培养箱和振荡器用于生产接种物，其中不使用孢子接种物，主要用于摇瓶实验。

（3）烤箱　烤箱可用于干燥玻璃器皿或测干重，或用作灭菌炉。烤箱工作室应由耐腐蚀不锈钢制成。大多数烤箱具有良好的空气循环，空气被吸入烘箱，通过加热元件，然后排出。

（4）泵　在实验室规模，几乎只通过持久泵实现液体泵送，隔膜计量泵也找到一些应用，但不太灵活。

（5）膜式过滤器　这些过滤器包含具有相同孔隙的醋酸纤维素膜/渗透性膜/聚四氟乙烯膜，因此截留大于该孔径的颗粒。规格为 $0.2\,\mu m$ 或 $0.45\,\mu m$ 孔径过滤器适用于大多数应用。

填充床型过滤器（深层过滤器）没有均匀的孔径，并且颗粒去除的机理倾向于更复杂。通常，过滤器壳体填充物有玻璃棉或非吸收性棉绒。这种过滤器易于压实，并且可以允许发生湍流。在一定情况下，压降的突然波动可能导致颗粒或包装材料的释放。

Further Reading
Stirred Tank Reactors

Bioreactors for Solid-state Fermentation A modified aerated STR (Stirred Tank Reactor), the Frings ace tor (Frings), is used in vinegar production, In this design, air is drawn in and distributed via a high-speed hollow-body turbin rotor, connected to an air suction pipe. The aerator is self-aspiration and so some pressed air is not required.

The STR is the most widely used bioreactor for aerated fermentations perhaps because of its reliability and flexibility. Nervelessness, operating and investment costs atrelatively high. In addition, problems exist in designing impellers for very large fermenters due to the length of the impeller shaft. In bioreactors without mechanical agitation, such as tower and loop bioreactors, aeration and mixing is achieved with high a through-puts.

In tower fermenters, air is introduced at the base of the fermenter and mixing is due to the rising bubbles so that the shear on the organism is minimal. Tower fermenters have been used to produce citric aicd, by pellets of A. niger and by strains of Candida guilliermondii, and also for production of vinegar, industrial alcohol, and beer. Because vertical mixing is relatively poor in tower fermenters, they can be operated as continuous systems, with bottom entry feed and top exit, with high biomass retention. In non-aerated tower fermenters, used for brewing or industrial alcohol production, biomass retention is maximized by use of fermenting yeasts having good Flocculating properctics.

Stirred tank bioreactors consist of a cylindrical vessel with a motor driven central **shaft** that supports one or more agitators. The shaft may enter through the top or the bottom of the reactor vessel. Microbial culture vessels are generally provide with four baffles projecting into the vessel from the walls to prevent swirling and vortexing of the fluid. The baffle width is 1/10 or 1/12 of the tank diameter. The aspect ratio (i.e. height-to-diameter ratio) of the vessel is 3~5, except in animal cell culture applications where aspect ratios do not normally exceed 2. Often, the animal cell culture vessels are unbaffled (especially small-scale reactors) to reduce turbulence that may damage the cells. The number of impellers depends on the aspect ratio. The bottom impeller is located at a distance about 1/3 of the tank diameter above the bottom of the tank.Additional impellers are spaced approximately 1.2 impeller diameter distance apart.The impeller diameter is about 1/3 of the vessel diameter for gas dispersion impellers such as Rushton disc **turbines** and concave bladed impellers. Larger hydrofoil impellers with diameters of 0.5 to 0.6 times the tank diameter are especially effective bulk mixers and gre used in fermenters for highly viscous mycelial broths. Animal cell culture vessesl typically employ a single, large diameter, low-shear impeller such as a marine propeller.

In animal or plant cell culture applications,the impeller speed generally does not exceed about 120rpm in vessels larger than about 50liters. Higher stirring rates are employed in microbial culture, except with mycelial and filamentous cultures where the impeller tip speed does not in general exceed 7.6m/s. Even lower speeds have been documented to damage certain **mycelial** fungi. The superficial aeration velocity in stirred vessels must remain below the value needed to flood the impeller. A flooded impeller is a poor mixer. Superficial aeration velocities do not generally exceed 0.05m/s.

New Words

shaft [ʃɑːft] n. 柄，轴；矛，箭；光线；vt. 给……装上杆柄

turbine ['tɜːbaɪn] n. 汽轮机；涡轮机；透平机

mycelial [maɪ'siːlɪəl] adj. 菌丝体的

参考文献

[1] Bisson L F. Stuck and sluggish fermentations. American Journal of Enology & Viticulture, 1999, 50(50):107-119.

[2] Bisson L F, Butzke C E. Diagnosis and rectification of stuck and sluggish fermentations. American Journal of Enology & Viticulture, 2000, 51(2):168-177.

[3] Zamora F. Biochemistry of Alcoholic Fermentation[M]//Wine Chemistry and Biochemistry. Springer New York, 2009:1037-1042.

[4] Ciani M, Beco L, Comitini F. Fermentation behaviour and metabolic interactions of multistarter wine yeast fermentations. International Journal of Food Microbiology, 2006, 108(2):239-245.

[5] Xu H, Miao X, et al. High quality biodiesel production from a microalga Chlorella protothecoides by heterotrophic growth in fermenters. Journal of Biotechnology, 2006, 126(4):499-507.

Chapter 5 Production of Antibiotics 抗生素

1. Antibiotics

Antibiotics have changed the world we live in. Their wide-scale introduction in the middle of the 20th century led to new standards of health for billions of people. Many of the life-threatening infections of previous centuries are now conveniently cured by oral medicine. Penicillin was the first major antibiotic from a microbial source to be commercialized. Its acceptance and success led to the search and identification of thousands of novel antibiotics. Many of which are now available for therapeutic use. Antibiotics also have applications as feed additives. Growth stimulants, **pesticides** and wider agricultural uses. The discovery of major antibiotics, such as penicillin, cephalosporin, **streptomycin, tetracycline** and erythromycins, and their subsequent development, have been well documented. Their commercial development over the past 50 years serves as an excellent example of how the applied research has contributed to producing low cost commodities that support therapeutic products with annual sales of the Multi-billions of dollars.

Antibiotics sold today are made either by total chemical synthesis or by a combination of microbial fermentation and subsequent chemical modification. The choice is one of simple economics. The microbial fermentation produces the basic active molecule at relatively low cost. And, through chemical modification. The **therapeutic** effects of the molecule can be

New Words and Expressions

pesticides ['pestɪsaɪdz]
n. 杀虫剂，除害药物

streptomycin ['streptə'maɪsɪn]
n. 链霉素
tetracycline [tetrə'saɪklɪn]
n. 四环素

therapeutic [θerə'pjuːtɪk]
adj. 治疗（学）的，疗法的；有益于健康的

increased. e. g. by increasing stability to low pH or temperature, widening the spectrum of activity, altering tissue distribution, increasing absorption and decreasing exeretion.

2. Penicillins

Penicillin and its related lactam cephalosporin, are bactericidal antibiotics. They inhibit the formation of peptide cross-linkages in the final stages of bacterial cell wall synthesis. Penicillin G and V are active against Gram-positive cocci but are readily inactivated by hydrolysis by m-lactamase-producing cultures and are therefore ineffective against staphylococcus aureus. **Cloxacillin** and floxacillin are resistant to-lactamase and are used against Staph. Aureus. The broad spectrum ampicillin and amoxicillin extend activity against Gram-negative bacteria such as Haemophilus influenza, *E.coli* and Proteus mirabilis. Amoxicillin is used in combination with clavulanic acid, a potent-lactamase inhibitor, to extend the use of this antibiotic, **Azlocillin** and **ticarcillin** are used to combat severe **pseudomonal** infections.

Penicillin G and V are fermented products from the fungus, penicillium chrysogenum. The bulk of penicillin G and V, however, is now used as starting material for the production of the active β-lactam nucleus, 6-aminpenicillinanic acid (6-APA).

Penicillin G can also be ring-expanded chemically to the cephalosporin nucleus which, after enzyme hydrolysis, yields the active nucleus 7-aminodeacetoxycehalosporanic acid (7-ADCA).

Both of these nuclei are important bulk products, and are used for the chemical synthesis

cloxacillin [klɒksəˈsɪlɪn]
n. 氯苯唑青霉钠；全霉林；邻氯西林；氯苯西林

azlocillin [eɪzləʊˈsɪlɪn]
n. 阿诺西林
ticarcillin [tɪkeəˈsɪlɪn]
n. α- 替卡西林，羟基噻吩青霉素
pseudomonal [sˈjuːdəʊməʊnl]
adj. 假单胞菌的

of the semisynthetic penicillins mentioned above, and in the case of 7-ADCA, for the synthesis of oral, broad spectrum **cephalosporins,** such as cefadroxil and cephalexin. The cephalosporins are particularly effective against urinary tract infections. The β-lactam nucleus can also be modified further to form the basis of antibiotics such as cefaclor and cefprozil.	cephalosporins [sefələus'pɔːrɪnz] *n.* 先锋霉素族抗生素
Cefadroxil and amoxicillin can now be made enzymically by using the reversible (synthetic) catalytic property of penicillin amidase. The penicillin G amidase from *E.coli* has been crystallized and its three-dimensional structure determined. Changing amino acids at the active site by site-directed mutagenesis has produced enzymes with improved ability to work in the synthetic direction.	
Addition of the precursor molecules, **phenylacetic acid** or **phenoxyacetic acid,** to fermentations of *p. chrysogenum* produce either penicillin G or penicillin V, respectively, for optimum production the culture is grown on a batch medium of corn steep liquor or soy flour plus minerals. And fed carbohydrate as a corn syrup throughout the cycle. In addition, the precursor and ammonium **sulphate** are fed to maintain critical concentrations of these components needed for the biosynthesis of penicillin.	phenylacetic acid ['fenələ'siːtik 'æsid] *n.* 苯乙酸 phenoxyacetic acid *n.* 苯氧基乙酸 sulphate ['sʌlfeɪt] *n.* 硫酸酯

参考译文

1. 抗生素

  抗生素改变了我们生活的世界。他们在20世纪中期的大规模引入为数十亿人带来了新的健康标准。许多以前几个世纪威胁生命的感染现在通过口服药物方便地治愈。青霉素是来自微生物来源的商业化的第一种主要抗生素。它的接受和成功导致成千上万的新型抗生素的发掘和鉴定。其中许多现在可用于治疗用途，

Unit 4　Fermentation Engineering　发酵工程

抗生素还具有作为饲料添加剂的应用，生长刺激剂，农药和更广泛的农业用途。主要抗生素如青霉素，头孢菌素，链霉素，四环素和红霉素的发现及其随后的发展已经被充分记录。他们在过去50年的商业发展是一个很好的例子，其应用研究上的说明能够帮助如何生产低成本商品，从而生产年销售额高达数十亿美元的治疗产品。

今天销售的抗生素可以通过全部化学合成或通过微生物发酵和随后的化学修饰的组合来制备。选择简单经济。微生物发酵以相对低的成本产生碱性活性分子，并通过化学改良。可以通过增加对低pH或温度的稳定性，拓宽活性谱，改变组织分布，增加吸收和减少兴奋来增加分子的治疗效果。

2. 青霉素

青霉素及其相关的内酰胺头孢菌素是杀菌性抗生素。它们在细菌细胞壁合成的最后阶段抑制肽交联的形成。青霉素G和V对革兰阳性球菌具有活性，但易于被产生m-内酰胺酶的培养物水解而失活，因此对金黄色葡萄球菌无效。氯唑西林和氟氯西林耐受β-内酰胺酶，并用于对抗金黄色葡萄球菌。广谱氨苄西林和阿莫西林扩展了对抗革兰阴性细菌如流感嗜血杆菌，大肠杆菌和奇异变形杆菌的活性。阿莫西林与克拉维酸（一种有效的内酰胺酶抑制剂）组合使用，以延长该抗生素的使用，阿洛西林和替卡西林用于对抗严重的假单胞菌感染。

青霉素G和青霉素V是来自真菌，青霉菌的发酵产物。然而，青霉素G和青霉素V现在大部分被用作生产活性β-内酰胺核6-氨基青霉烷酸（6-APA）的起始材料。

青霉素G也可以被化学法环化扩展成头孢菌素核，其在酶水解后产生活性核7-氨基脱乙酰氧基头孢烷酸（7-ADCA）。

这两个都是重要的产物，并且用于上述半合成青霉素的化学合成，在7-ADCA的情况下，用于合成口服的广谱头孢菌素，例如头孢羟氨苄和头孢氨苄。头孢菌素对尿道感染特别有效，β-内酰胺核还可以进一步修饰以形成抗生素如头孢克洛和头孢丙烯的基础。

现在可以通过使用青霉素酰胺酶的可逆（合成）催化性质，利用酶法制备头孢羟氨苄和阿莫西林。来自大肠杆菌的青霉素G酰胺酶已经结晶并确定其三维结构。在活性位点通过位点-定向诱变产生具有在合成方向上改进工作能力的酶。

将前体分子苯乙酸或苯氧基乙酸添加到产黄青霉的发酵中分别产生青霉素G或青霉素V。为了最佳生产，将培养物在玉米浆或大豆粉加矿物质的分批培养基上生长。并在整个循环中将玉米糖浆加入碳水化合物。此外，加入前体和硫酸铵以保持青霉素生物合成所需组分的临界浓度。

Further Reading

Cephalosporins

Cephalosporins were developed to overcome the allergic problems associated with penicillins. They can, however, be modified chemically at two sites: the 7-amino and the 3-methylene, to produce a variety of very effective antibiotics, notably cephamandole, cefazolin and cefepime.

Cephalosporins are made from cephalosporin C, a fermented product of Acremonium chrysogenum which, after extraction and purification, is hydrolysed, either enzymically or chemically, to the active nucleus, 7-aminocephalosporanic acid (7-ACA), which serves as substrate for the chemical synthesis of injectable, semisynthetic cephalosporins, Cephalosporins with a 7-α-methoxy group (cephamycins) are produced by several Streptomyces spp. and they serve as the precursor of **cefoxitin a**nd others.

High-producting cultures of *A.chrysogenum* are fermented using corn steep liquor and soy flour-based media with continuous feeding of both corn syrup and triacylglycerols (soy, rape or lard oil). Methionine is used both as a source of Sulphur and as an inducer of morphological changes. Cephalosporin C is unstable and degrades chemically to diacetyl Cephaloporin C (DAC) and a thiazole-4-carboxylate, Cephalosporin C is also hydrolysed to DAC by esterases released by the fungus.These, however, can be inhibited by the use of phosphates.

Aminoglycosides

Streptomycin was the first aminoglycoside used for antibiobic therapy. Its activity against Mycobacterium tuberculosis initiated the wide-spread introduction of antibiotic treatment to combat tuberculosis. Aminoglycosides are potent antibiobics and have activity against both Gram-positive and Gram-negative bacteria as well as against mucobacteria. Unfortunately, they can have nephron-(kidney) and oto-toxicities (hearing), and care has to be taken in their use in treatment of serious **infections.**

Aminoglycosides are bactericidal and work by binding to the 30s ribosome subunit which prevents protein synthesis. There are many aminoglycosides in medical use and are all derived from actinomyces spp. For example, streptomycin. Gentamicin, tobramycin, kanamycin, sisomicin.

Large-scale fermentations of aminoglycosides have several similar features. Use of soy products is common, e. g. soy flour or soy meal.Antibiotic synthesis is sensitive to feedback repression by glucose , ammonia and phosphate.For these reasons ammonium and phosphate salts are not used in the starting batch.Nitrogen is obtained from the slow metabolism of the soy proteins and the necessary phosphate is obtained from organic sources such as phytic acid. Starch is commonly used in the starting batch as Streptomyces have poor amylase activities and the enzymic release of glucose is slow and rate limiting.Alternatively, corn syrups can be fed at pre-determined rates.

New Words

cefoxitin ['sefɒksɪtɪn] n. 头孢西丁，头孢甲氧霉素

infections [ɪn'fekʃnz] n. 传染病

参考文献

[1] Chen X H, Koumoutsi A, et al. More than anticipated-production of antibiotics and other secondary metabolites by Bacillus amyloliquefaciens FZB42. Journal of Molecular Microbiology & Biotechnology, 2008, 16(1-2):14-24.

[2] I.S. Hornsey, D. Hide. The production of antimicrobial compounds by British marine algae II. Seasonal variation in production of antibiotics. European Journal of Phycology, 1976, 11(1):63-67.

[3] Homma Y, Sato Z, et al. Production of antibiotics by Pseudomonas cepacia as an agent for biological control of soilborne plant pathogens. Soil Biology & Biochemistry, 1989, 21(5):723-728.

Unit 5　Enzyme　酶工程

Chapter 1　The Biological Catalysts of Life　生命的生物催化剂

1. Introduction

Enzymes are proteins functioning as catalysts that speeds up reactions by lowing the activation energy. A simple and **succinct** definition of an enzyme is that it is a biological catalyst that accelerates a chemical reaction without altering is its **equilibrium**. During the reactions the enzymes themselves undergo transient changes. In the overall process, enzymes do not undergo any net change. The enzyme catalysts regulate the structure and function of cells and organisms. They catalyze the synthesis and breakdown of biochemical building blocks and **macromolecules**, the transmission of genetic information, the transport of compounds across the membranes, **motility** of organisms and conversion of chemical energy. Enzyme catalysis is essential for making biochemical reactions proceed at appropriate speed in physiological conditions. They speed up the reactions in the cells so that they may occur in fractions of seconds. In the absence of catalysts most cellular reactions would not occur even time periods of years. Without rapid cellular reactions life in its present form would not be possible. One characteristic feature of enzymes is their **specificity**. Thus each reaction in the cell is catalyzed by its own, specific enzyme. The substances that are acted upon by enzymes are substrates. Substrate can be a small molecule or

New Words and Expressions

enzyme [ˈenzaɪm]
n. 酶

succinct [səkˈsɪŋkt]
adj. 简洁的，简明的

equilibrium [ˌiːkwɪˈlɪbriəm]
n. 平衡点

macromolecule [ˌmækrəˈmɒləkjuːl]
n. 高分子

motility [məʊˈtɪlɪtɪ]
n. 运动性

specificity [ˌspesɪˈfɪsəti]
n. 专一性，特异性

macromolecule like an enzyme itself. For example, **trypsin** is the enzyme that uses **polypeptides** as the substrate and **hydrolyses** the peptide bonds.

The exact number of different enzymes in various cells is not known. However, because the number of different reactions in the cells of higher **eukaryotes** is in the tens of thousands, the number of different enzymes in the cells has to be in the same scale. Based on results of the **genome sequencing** projects, the estimated number of enzymes in the cells is now much more accurate than before these sequencing projects were initiated. For example, in an *Escherichia coli* cell, there are roughly 4300 proteins and almost 3000 of them are enzymes. **Mammals** have more than ten times the number of proteins and enzymes than there are in *E. coli*.

Enzymes have been utilized for thousands of years in microbial processes. Microbes and their enzymes have been applied for preparation of wines, beer, cheeses and other milk products. The role of enzymes in the fermentation process has been known for less than two hundred years. In the 1850s Lous Pasteur presented a theory that sugar is converted into ethanol in yeast by "ferments". He also concluded that these ferments are inseparable from the living yeast cells. At the end of the eighteenth century, more information on the nature of fermentation was obtained when Buchner was able to ferment using a yeast filtrate. Processes for this isolation of enzymes and studies of their properties were ready to start.

The **ultracentrifugation** technique was applied for enzymes studies in the 1920s. These studies revealed the physical nature of the enzymes. The 3-dimensional structure of **lysozyme** was solved in

trypsin [ˈtrɪpsɪn]
n. 胰蛋白酶
polypeptide [ˌpɒlɪˈpeptaɪd]
n. 多肽
hydrolyse [ˈhaɪdrəlaɪz]
vi. 水解
eukaryote [juˈkærɪəʊt]
n. 真核细胞
genome sequencing
[ˈdʒiːnəʊm ˈsiːkwənsɪŋ]
n. 基因组测序

Escherichia coli
[ˌeʃəˈrikiə ˈkəʊlaɪ]
n. 大肠杆菌
mammal [ˈmæml]
n. 哺乳动物

ultracentrifugation
[ˌʌltrəsentrɪfjʊˈgeɪʃn]
n. 超速离心
lysozyme [ˈlaɪsəzaɪm]
n. 溶菌酶

1955 by using **X-ray diffraction** analysis. On the basis of the 3-dimensional structure, the significance of the active in the action mechanism was postulated. Today thousands of amino acid sequence of proteins are known. Equally, the number of known 3-dimensional protein structures is thousands. Most proteins have tight globular structure, and they contain one or several **subunits.** However, enzymes show considerable flexibility. Some are regulated by small molecules (**effectors**). In the cells there are small molecule, often end products of the enzyme reaction, which regulate the level of enzyme activity. This is called **allosteric regulation**.

Development of **recombinant** DNA technology in the 1970s had enormous impact in understanding of protein structure/function relationship. By applying gene manipulation, it was possible to study in a rational manner amino acid residues involved in protein stability, substrate binding, and enzyme catalysis and subunit interaction. It also facilitated in developing methods for protein purification and even opened new views for designing proteins with desired structure and properties. By **site directed mutagenesis**, certain amino acid residues were changed in the active center resulting in change in substrate specificity and catalysis.

Most biological catalysts are proteins, but not all. RNA can also act as catalyst in RNA hydrolysis speeding up the reaction up to 10^{11} fold. These catalysts are called **ribozyme**s.

2. Classification and nomenclature of enzyme

Enzymes are typically classified according to the types of reactions they catalyze. In the Enzyme

X-ray diffraction
[eksrei di'frækʃən]
n. X-射线衍射

subunit [sʌb'juːnɪt]
n. 亚基
effector [ɪ'fektər]
n. 效应物

allosteric regulation
[ælə'sterik ˌregju'leiʃən]
n. 变构调节
recombinant [rɪ'kɒmbɪnənt]
adj. 重组的

site directed mutagenesis
[saɪt dɪ'rektɪd ˌmjuːtə'dʒenɪsɪs]
n. 定向诱变

ribozyme ['raɪbəˌzaɪm]
n. 核酶

Nomenclature classification, they are subdivided and categorized into six main enzyme classes corresponding to the type of reactions such as enzyme catalyze. Table 5-1 gives an overview of this categorization, in particular the main enzyme classes.

Table 5-1 The main enzyme classes

Enzyme class	Catalyzed reaction
1. **Oxidoreductases**	Oxidation-reduction reactions
2. **Transferases**	Transfer of functional groups
3. **Hydrolases**	Hydrolysis reactions
4. **Lyases**	Group elimination (forming double bonds)
5. **Isomerases**	Isomerization reactions
6. **Ligases**	Bond formation coupled with a triphosphate cleavage

The Enzyme Nomenclature suggests two names for each enzyme, a **recommended name** convenient for everyday use and a **systematic name** used to minimize ambiguity. Both names are based on the nature of the catalyzed reaction. The recommended name is often the former trivial name, sometimes after little change to prevent **misinterpretation**.

As the systematic name may be very extensive nd uncomfortable to use, the **Enzyme Commission** (EC) has also developed a numeric system based on the same criteria, which can be used together with the recommended name to specify the mentioned enzyme. According to this system, each enzyme is assigned a four-digit EC number (table 5-2). The first three numbers define major class, **subclass** and **sub-subclass**, respectively. The first number indicates the type of reaction. The second number specifies the bond cleaved or the type of the substrate.

nomenclature [nəˈmenklətʃər]
n. 系统命名法
oxidoreductase [ˈɒksɪdəʊrɪˈdʌkteɪs]
n. 氧化还原酶
transferase [ˈtrænsfəreɪs]
n. 转移酶
hydrolase [ˈhaɪdrəleɪs]
n. 水解酶
lyase [ˈlaɪəs]
n. 裂合酶，裂解酶
isomerase [aɪˈsɒməreɪs]
n. 异构酶
ligase [lɪˈgeɪs]
n. 连接酶
triphosphate [traɪˈfɒsfeɪt]
n. 三磷酸盐
cleavage [ˈkliːvɪdʒ]
n. 分裂，裂解
recommended name [rekəˈmendɪd neɪm]
n. 习惯名称
systematic name [sɪstəˈmætɪk neɪm]
n. 系统名称
misinterpretation [ˈmɪsɪnˌtɜːprɪˈteɪʃən]
n. 误解
Enzyme Commission [ˈenzaɪm kəˈmɪʃən]
n. 酶学委员会
subclass [ˈsʌbklˈɑːs]
n. 亚类
sub-subclass [ˈsʌb ˈsʌbklˈɑːs]
n. 亚亚类

The third number defines more precisely the catalyzed reaction. The fourth number indicates the serial number of the enzyme in its sub-subclass.

Table 5-2　Constitution of the four-digit EC number

EC number EC(i).(ii).(iii).(iv)

(i) the main class, denotes the type of catalyzed reaction

(ii) sub-class, indicates the substrate type, the type of transferred functional group or the nature of one specific bond involved in the catalyzed reaction

(iii) sub-subclass, expresses the nature of substrate or co-substrate

(iv) an arbitrary serial number

3. Enzyme applications

Application of enzymes in different industries is continuously increasing during the last two decades. Applications of enzymes in food industries include **baking**, **dairy** products, **starch** conversion and **beverage** processing (beer, wine, fruit and vegetable juices). In textiles industries, enzymes have found a special place due to their effect on end products. In industries such as pulp and paper making and **detergents**, the use of enzymes has become an inevitable processing strategy when a perfect and product is desired. Application of enzymes in more modern industries including biosensors is improving rapidly due to the specificity of enzyme which is of prime importance in biosensor. Many other important industries including health care and **pharmaceuticals** and chemical manufacture are increasing taking advantages of these natures amazing catalysts.

Enzymes have found a wide spectrum of applications, from a very large number of enzyme-

baking ['beɪkɪŋ]
n. 焙烤
dairy ['deəri]
adj. 乳品的，牛乳的
starch [stɑːtʃ]
n. 淀粉
beverage ['bevərɪdʒ]
n. 饮料
detergent [dɪ'tɜːdʒənt]
n. 洗涤剂，去垢剂

pharmaceuticals [ˌfɑːmə'sjuːtɪklz]
n. 医药品，药物

catalyzed industrial processes to use within the toolbox of molecular biology. Besides their use as a process catalysis in industrial processes. Enzymes have found important applications in:

(1) Chemical and clinical analysis, due to their high specificity and sensibility, which allow the quantification of various **analytes** with high precision. Enzymes are also widely used in various diagnostic kits and as detectors in **immunoassays**.

analyte [ænə'lɪt]
n. 分析物
immunoassay [ˌɪmjʊnəʊ'æseɪ]
n. 免疫法

(2) Therapy, due to their high specificity and activity, which allow the precise and efficient removal of unwanted metabolites. Many therapeutic applications have been envisaged. The US Food and Drug Administration (FDA) has already approved applications in **cardiovascular** disorders, **pancreatic insufficiency**, several types of cancer, the replacement therapy for genetic deficiencies, the **debridement** of wounds, and the removal of various toxic metabolites from the bloodstream.

cardiovascular [ˌkɑːdiəʊ'væskjələr]
adj. 心血管的
pancreatic insufficiency [ˌpæŋkrɪ'ætɪk ˌɪnsə'fɪʃənsɪ]
n. 胰腺机能不全
debridement [dɪ'brɪdmənt]
n. 清创术

(3) Environmental management (waste treatment and **bioremediation**). Enzyme specificity allows the removal of particularly recalcitrant pollutants from hard industrial wastes and highly diluted effluents. In addition, they are increasingly being used in bioremediation of polluted soils and waters with recalcitrant compounds, although the high cost of the enzymes is still prohibitive in most such cases.

bioremediation [baɪərmiːd'ieɪʃn]
n. 生物修复

(4) Biotechnological research and development. Enzymes are fundamental components of the toolbox for biotechnology research, especially in the areas of molecular biology and genetic engineering. Thermostable DNA **polymerases** are the basis of the polymerase **endonucleases** and ligases are fundamental tools in recombinant DNA technology.

polymerase ['pɒlɪməreɪs]
n. 聚合酶
endonuclease [endə'njuːklɪeɪs]
n. 核酸内切酶

Highly purified enzymes are required for most of these applications, which are sold at very high unitary prices. Therefore, despite their low volume of production, the market size is significant.

参考译文

1. 简介

酶是一种功能性蛋白质，作为生物催化剂，通过降低反应活化能来加快反应速度。酶的简单定义，它是一种加速化学反应而不改变反应平衡的生物催化剂。在催化反应过程中，酶发生瞬态变化。酶在整个反应过程中本身并不消耗。酶催化调节细胞和生物的结构和功能。酶可以催化大分子的合成和分解，遗传信息的传递，物质的跨膜转运，生物体的运动和化学能的转化，这些都有酶的参与。酶催化是在生理条件下进行生物化学反应的关键，它们加快细胞的反应，从而使反应在几秒钟内发生。在没有催化剂的情况下，甚至经过了数年都不会发生化学反应。如果生命细胞中没有快速的反应，生命就不可能以现在的样子出现。酶的特征之一是特异性，因此，在细胞中每个反应是由其特定的酶来催化的。酶催化作用的物质称为底物。催化底物可以是小分子物质，也可以是类似酶的大分子物质。例如，胰蛋白酶是以多肽为底物，水解肽链的酶。

不同细胞中酶的具体数量是不知道的。然而，由于在高等真核生物细胞中发生的反应数量是成千上万的，细胞中酶的数量规模与反应数量规模是相当的。根据基因组测序的结果，对细胞中酶的数量估计比以前更为准确。例如，在大肠杆菌细胞中，大约有4300种蛋白质，其中几乎3000种都是酶。哺乳动物的蛋白质和酶的数量是大肠杆菌的十倍。

在微生物生产过程中酶的使用已经达到数千年。微生物及其酶可用于制备葡萄酒、啤酒、干酪等乳制品。人们发现酶在发酵过程中所起的作用不到200年。在19世纪50年代，巴斯德提出了一个理论，在酵母细胞中糖通过"发酵"转化为乙醇。他还认为，发酵是离不开活的酵母细胞。在18世纪末，当Bücher用酵母滤液发酵的时候，他获得更多发酵本质的信息。这已经开启了酶的分离和特性的研究。

在20世纪20年代，超速离心技术开始用于酶的研究。这些研究揭示了酶的理化性质。1965年，通过X-射线衍射分析法获得溶菌酶的三维结构。在三维结构的基础上，推断酶作用机制中活性的重要性。今天已经知道成千上万蛋白质的氨基酸序列，同样，已知的蛋白质的三维结构也是成千上万的。大多数蛋白质有紧密的球状结构，它们包含一个或几个亚基。然而，酶具有相当大的灵活性。某些酶受到小分子（效应物）的调节，在细胞中这些效应物是小分子物质，通常是酶催化反应的末端产物，它调节酶的活性水平，这被称为变构调节。

20 世纪 70 年代，重组 DNA 技术的发展，对蛋白质结构/功能关系的分析有了巨大的影响。通过基因操作，可以用理性的方式研究氨基酸残基，包括蛋白质的稳定性、底物结合、酶催化和亚基间相互作用。这也促进了蛋白质纯化方法的发展，甚至开辟了设计所需结构和性能蛋白的新视野。通过位点定向诱变，改变活性中心的某些氨基酸残基，从而导致底物特异性和催化活性发生变化。

大多数的催化剂是蛋白质，但不是全部。RNA 也可以作为催化剂，它可催化 RNA 的水解，催化速率可达到 10^{11} 倍。这些催化剂被称为核酶。

2. 酶的分类与命名

酶通常根据催化反应类型进行分类。在酶的命名和分类中，酶被分为六类。表 5-1 给出了酶的分类。

表 5-1　　　　　　　　　　酶的主要分类

酶的分类	催化反应
1. 氧化还原酶类	氧化还原反应
2. 转移酶类	功能基团的转移
3. 水解酶类	水解反应
4. 裂解酶类	消除一个基团，形成双键
5. 异构酶类	异构化反应
6. 连接酶	ATP 参与下形成连接键

酶命名建议有两个名字，一个是习惯名称可以方便于日常的使用，另一个是系统名称来用于减少歧义。两个名称都是基于催化反应性质，推荐名称往往是系统名称的俗称，以防止误解，有时变化不大。

由于系统名称冗长和使用不方便，酶学委员会（简称为 EC）开发了国际系统分类法，采用相应的数字指定所提到的酶。根据这个系统，每个酶具有的四个数字 EC 编号，见表 5-2。第一个数字表示酶催化反应类型。根据底物性质、转移基团的类型或化学键的性质，将催化反应类型继续细分为亚类，第二个数字代表亚类。第三个数字是亚亚类，是根据底物或共底物对亚类的进一步细分。第四个数字代表具体酶的编号。

表 5-2　　　　　　　　　　国际系统编号

EC (ⅰ).(ⅱ).(ⅲ).(ⅳ)
(ⅰ) 大类，指的是酶催化反应类型
(ⅱ) 亚类，指底物类型、转移基团的类型或化学键的性质
(ⅲ) 亚亚类，底物或共底物的性质
(ⅳ) 酶的具体编号

3. 酶的应用

近20年以来，酶在不同行业中的应用不断增加。酶在食品工业中的应用主要包括烘焙、乳制品、淀粉糖和饮料加工（啤酒、葡萄酒、果汁和蔬菜汁）。在纺织行业中，酶的特殊作用对纺织产品产生影响。在纸浆、造纸和洗涤剂等行业中，为了使产品达到预期的质量，在产品加工中酶的应用已成为一种必然的手段。酶在现代工业中的应用，包括利用酶的专一性提高生物传感器的检测灵敏度，对于生物传感器来说，这是最重要的性能。许多其他重要的行业包括医疗保健、药品和化学药品制造业正在不断增加酶的使用。

酶的使用范畴非常广泛，从大规模的酶催化工业产品到分子生物学的工具酶，除了在工业产品生产中的应用外，还有其他重要的应用：

（1）化学和临床分析，由于其高特异性和敏感性，可以对不同的分析物进行高精度的定量分析。酶也广泛用于各种诊断试剂盒和免疫测定中探测器。

（2）治疗，基于酶的高特异性和活性，因此可以精确和有效地清除有害的代谢物。许多酶在治疗疾病方面已有应用，包括天冬酰胺酶、胆红素氧化酶和尿激酶等。美国食品和药品管理局（FDA）已经批准酶在心血管疾病、胰腺功能不全、几种类型的癌症、遗传缺陷的替代疗法、伤口清理和血液中的各种有毒代谢物清除方面的应用。

（3）环境管理（废物处理和生物修复）。利用酶的特异性去除非常难以降解的污染物，来源于工业废水，特别是高度稀释的废水。此外，酶也越来越多地用于土壤污染和水污染的生物修复。但是由于酶的成本高，在大多数情况下仍然限制使用。

（4）生物技术的研究与开发。酶是生物技术研究工具的基本组成部分，特别是在分子生物学和基因工程领域。耐热DNA聚合酶和连接酶是DNA重组技术的基本工具。

高纯度酶的需求主要在以上几个方面，这些酶的统一售价是非常高的。因此，尽管高纯度酶的生产规模小，但市场规模是显著的。

Further Reading
Biofuel Speeds Ahead with Enzyme Technology

New advances in the use of enzymes in **biofuels**, particularly in cellulosic ethanol production, have enabled the biofuel industry to keep its innovative edge. **Cellulosic ethanol** shows enormous potential, in terms of both output volumes and environmental saving- allowing for up to 90% of carbon dioxide emissions reduction compared with **petroleum**-based fuels. In the USA, federal regulations compelling 16 bn gallons of cellulosic ethanol use by 2022 reflect such well-recognized potential of this advanced biofuel. A broad variety of low-cost, readily available **feedstocks**, such as **sugarcane bagasse** and municipal solid wastes, can be used to produce cellulosic ethanol.

Several companies, such as Novozymes, have successfully developed enzyme-based technologies to convert biomass into ethanol at a commercially feasible cost. The scale-up by ethanol producers of their cellulosic processes from laboratory level to pilot, and even to demonstration, levels have been made possible by the most advanced cellulosic enzymes, such as Novozymes **CellicCTEC2**. The enzyme technology is commercially available, promising improved operating costs and reduced capital requirements. Novozymes's enzymes also found their way at piedmont Biofuels' new pilot plant in the USA, which was launched in 2010 to demonstrate an enhanced technological process for biodiesel production. The plant makes use of advanced technology and Novozymes's enzymes to make premium **biodiesel** (FAME) with low-quality waste grease as feedstock. The new process is claimed to boost yields, reduce waste, and allow the use of lower-cost raw materials. The plant, expected to have an initial production of 12,600 gallon/y of biodiesel, is the result of collaboration with the Biofuels Center of North Carolina and the Chatham Country Economic Development Corp.

New Words

biofuel ['baɪəʊ'fjuːəl] *n.* 生物燃料

cellulosic [seljʊ'ləʊsɪk] *adj.* 有纤维质的

petroleum [pɪ'trəʊlɪəm] *n.* 石油

feedstock ['fiːdstɒk] *n.* 原料；给料（指供送入机器或加工厂的原料）

sugarcane ['ʃugə,keɪn] *n.* 甘蔗；糖蔗

bagasse [bə'gæs] *n.* 甘蔗渣

CellicCTEC2 复合纤维素酶的一种

biodiesel ['baɪəʊdiːzl] *n.* 生物柴油；生物质柴油

参考文献

[1] Jan B van, Beilen, et al. Enzyme technology: an overview. Current Opinion Biotechnology. 2002, 13:338-344.

[2] Andrés Illanes, Lorena Wilson, et al. Problem solving in Enzyme Biocatalysis. John Wiley & Sons, 2013.

[3] Drauz, Karlheinz et al. Enzyme Catalysis in organic Synthesis. Wiley-VCH Verlag GmbH, 2012.

[4] Relx Group. Biofuel speeds ahead with enzyme technology. Oils and Fats International, 2011, 27(2):23-24.

[5] Andreas Liese, Karsten Seelbach, et al. Industrial biotransformations. Wiley-VCH Verlag GmbH, 2006.

Chapter 2　Enzyme Production　酶的生产

1. Summary

Commercial enzyme production has grown during the past century in volume and number of product in response to expanding market and increasing demand for novel biocatalysts. Microorganisms constitute the major source of enzymes. Traditional enzyme production relied on the natural hosts as raw materials, however genetic engineering has now given a choice for producing sufficient quantities of enzymes in selected production **hosts** microorganisms and **transgenic** plants.

Production of a new microbial enzyme starts with **screening** of microorganisms for desirable activity using appropriate selection procedures. The level of enzyme activity produced by an organism from a natural environment is often low and needs to be elevated for industrial production. Increase in enzyme level is often achieved by mutation of the organism. An alternative strategy that has gained favor is production of the enzyme in a recombinant organism of choice whose growth conditions are well optimized and whose **GRAS** (generally regarded as safe) status is established. **Random** or site-directed **mutagenesis** with the purpose of engineering the activity and stability properties of an enzyme prior to its production is becoming a common practice. The microorganisms used for enzyme production are grown in fermenters using an optimized growth medium. Both solid state- and submerged fermentation are applied commercially, however the latter is preferred in many countries because of a better handle on aseptic conditions

New Words and Expressions

host [həʊst]
n. 寄主
transgenic [ˌtrænz'dʒenɪk]
adj. 转基因的，基因改造的
screening ['skriːnɪŋ]
n. 筛选

GRAS (generally recognized as safe) Abbr. 一般认为安全（美国食品及药物管理局用语）
random
['rændəm]
adj. 随机
mutagenesis
[ˌmjuːtə'dʒenisis]
n. 诱变

and process control. The enzymes produced by the microorganism may be **intracellular** or secreted into the **extracellular** medium.

Isolation and purification, i.e., downstream processing of enzyme from the raw material constitutes the subsequent key stage in the production process. The desired level of purification depends on the ultimate application of the enzyme product. The industrial bulk enzymes are relatively crude formulations while speciality enzymes undergo a thorough purification to yield a **homogeneous** product. A traditional downstream processing scheme involves stages of clarification for separation of the enzyme from the solids comprising the raw material, concentration to reduce the process volumes, and purification to separate it from other soluble contaminants. In case of the intracellular enzymes, disruption of cells or tissue for release of the product is among the primary separation steps. There is a choice of different separation techniques for each stage. **Chromatography** is the major technique for **high-resolution** purification of enzymes. Some separation techniques allow integration of the downstream processing stages required for purification thus reducing the number of steps and hence the production costs. The enzyme is finally formulated as a liquid or solid product. In either case, stabilizing additives are added for rendering long shelf life to the product. Some enzymes are immobilized to solid supports or enzyme crystals are **cross-linked** to render them insoluble and stable for repeated or long term use in a process application.

2. Enzyme sources

The primary consideration in the production of

---

intracellular [ˌɪntrəˈseljʊlə] *adj.* 细胞内的

extracellular [ˌekstrəˈseljʊlə] *adj.* （位于或发生于）细胞外的

homogeneous [ˌhɒməˈdʒiːniəs] *adj.* 同质的，同种的

chromatography [ˌkrəʊməˈtɒɡrəfɪ] *n.* 色谱，层析

high-resolution [ˈhaiˌrezəˈluːʃən] *n.* 高分辨率

cross-linked [krɔs liŋkt] *vi.* 交联

any enzyme relates to the choice of source. In most cases, the desired activity can be obtained from several sources. Traditionally, however, the choice of source has been more restricted for some enzymes. For example, the enzyme **rennet** was until recently obtained from the stomach of **suckling calves**; the corresponding microbial enzyme led to an **off-flavor** in the cheese produced.

Microorganisms represent an attractive source of enzymes as they can be cultured in large quantities in relatively short period by established methods of fermentation. However, the level of production of a particular enzyme varies in different microorganisms, and moreover the enzymes often differ in composition and properties. One usually finds the closely related organisms have enzymes with nearly similar properties, while unrelated organisms have enzyme systems that differ widely. The most critical feature of the organisms for producing industrially significant enzymes is their GRAS status, which implies that they must be non-toxic, **non-pathogenic** and generally should not produce antibiotics.

The GRAS listed microorganisms included fewer than 50 bacteria and fungi. Examples are the bacteria including ***Bacillus subtilis***, *B licheniformis*, and various other bacilli, ***lactobacilli***, ***Streptomyces*** species, the yeast ***Saccharomyces cerevisiae***, and the filamentous fungi, etc. In case of Bacillus, mutants are selected that can no longer form spores.

Since ***Aspergillus*** cultures are frequently inoculated with **conidia**, enzyme production using these fungi relies on good spore formation. Most of the bulk enzymes (hydrolases) are secreted by the microorganisms directly into the culture medium,

rennet ['renɪt]
n. 凝乳酵素，凝乳酶
suckling calves ['sʌklɪŋ kɑːvz]
n. 哺乳犊牛
off-flavor ['ɔːffl'eɪvər]
n. 异味

non-pathogenic ['nɒnpæθədʒ'enɪk]
adj. 非病源的，不致病的
Bacillus subtilis [bə'sɪləs 'sʌbtɪlɪs]
n. 枯草芽孢杆菌
lactobacilli [læktəbəsɪ'liː]
n. 乳酸杆菌
Streptomyces [ˌstreptəu'maɪsiːz]
n. 链霉菌
Saccharomyces cerevisiae [sækərəu'maɪsiːz sɪrɪ'vɪziː]
n. 酿酒酵母
Aspergillus [ˌæspə'dʒɪləs]
n. 曲霉菌
conidia [kəu'nɪdɪə]
n. 无性孢子，分生孢子

while some enzymes e,g **penicillin acylase** and glucose isomerase are intracellular. For some applications, it may not be necessary to isolate the enzyme but the microbial cells themselves are used as enzyme source.

  The organism is preferred which gives high yield of enzyme in shortest possible fermentation time. The production strains used in industry are normally modified by genetic manipulation to have high levels of production.

  A common trend in the industry today is that the gene coding for enzyme with desired characteristics is transferred into one of the selected microbial production strains which have all the required features of safety and high expression levels and for which the growth medium has been optimized, hence avoiding the need for optimization of individual enzyme producing strains.

  Despite the advantages of microorganism as enzyme source, some enzymes are still economically produced from plant and animal sources. This is possible because of sufficiently high amounts of these enzymes in such sources and also as a means to convert inexpensive, renewable material like agricultural and **slaughter** waste into value added products.

3. Strain development

  When a prospective microorganism producing a desired enzyme activity has been identified, several steps are required to convert it into a strain that is suitable for commercial use. The wild type organism invariably produces the desired enzyme in amounts too low for commercial use. One major **impediment** to high levels of enzyme production is the phenomenon of **catabolite repression**, in

---

penicillin acylase [ˌpenɪˈsɪlɪn ˈæsɪleɪs]
n. 青霉素酰化酶

slaughter [ˈslɔːtər]
n. 屠宰

impediment [ɪmˈpedɪmənt]
n. 妨碍，阻止

catabolize repression [kəˈtæbəlait riˈpreʃən]
n. 分解代谢物阻遏

which the presence of readily metabolized carbon source such as glucose represses the biosynthesis of a number of degradative enzymes. Another problem is that degradative enzymes are produced in significant quantity only when a specific inducer is present in the medium. Some **inducers** are too expensive to be used in large-scale enzyme production.

inducer [ɪnˈdjuːsə]
n. 诱导物

For industrial purposes, the organism should grow on an inexpensive medium and give a constant, high yield of enzyme in a short time. Secondary enzyme activities and the content of metabolites in the fermentation broth should be minimal which would facilitate simple recovery of the enzyme. Lastly, the process must be safe to the personnel in the production plant, and the **effluents** from the plant should not disturb the environment. The fulfilment of these objectives requires a combined optimization of the strain properties and process parameters.

effluent [ˈefluənt]
n. (注入河里等的) 污水, 工业废水

Optimization of strain properties, referred to as strain development is very attractive because this, as a rule, offers an inexpensive and permanent solution to most of the limitations mentioned above, i.e.:

(1) Yield and/or productivity of the enzyme can be increased,

(2) Contaminant enzymes can be reduced in number or eliminated,

(3) Metabolism of the producer organism can be altered so that the desired product is produced constitutively, and

(4) The organism can be modified to produce the enzyme under conditions which would normally be catabolize repressive or **end-product inhibitory**. Furthermore, specificity of an enzyme

end-product inhibitory
[ˈendprˈɒdʌkt ɪnˈhɪbɪtərɪ]
n. 末端产物抑制

can be altered, activity and stability of the enzyme under specific conditions can be increased, and also health and safety problems can often be avoided by strain development.

As the productivity and metabolic functions of the organisms are genetically controlled, the strains development is achieved by changes at the gene level combined with a suitable selection mechanism. There are three different approaches available for strain improvement. These include mutation, hybridization and recombinant DNA technology.

4. Growth requirements of microorganisms

Cultivation of the host microorganism is the actual enzyme production step. The growth medium for microorganisms is designed to provide the essential elements. Optimization of the medium is important for desirable yield and cost of an enzyme preparation.

At industrial scale, it is important that the producer microorganism is able to grow rapidly on a cheap medium since substrate expenses can easily account for up to 80 percent of the cost of the fermentation process. Carbon is often supplied as a carbohydrate. The typical carbohydrate constituents of inexpensive media are **molasses** as a source of sucrose; **barley**, corn, wheat, starch **hydrolysate**s, etc. to provide glucose, and **whey** as a source of lactose. Nitrogen can be provided as simple inorganic salts (**ammonium** and **nitrate** salts) or complex materials such as **soybean meal**, **corn steep liquor**, **peptone**, etc. Appropriate inorganic salts provide supplementary sources of sulfur, calcium and etc. Trace metals are contributed either by the **tap water** used for preparation of the media or are supplemented to the medium. Some

molasses [məˈlæsɪz]
n. 糖蜜
barley [ˈbɑːli]
n. 大麦
hydrolysate [haɪˈdrɒlɪseɪt]
n. 水解产物
whey [weɪ]
n. 乳清
ammonium [əˈməʊniəm]
n. 铵
nitrate [ˈnaɪtreɪt]
n. 硝酸盐
soybean meal [ˈsɔɪˌbiːn miːl]
n. 大豆粉，豆粕
corn steep liquor [kɔːn stiːp ˈlɪkə]
n. 玉米浆
peptone [ˈpeptəʊn]
n. 蛋白胨
tap water [tæp ˈwɔːtə]
n. 自来水

organisms also require specific growth factors such as vitamins. Incorporation of low concentrations of a **surface-active agent** like Tween 80 in the medium has been used to improve the yield of extracellular enzymes by way of facilitating release of the enzymes from cells.

An **inducer** if required could be a substrate or a structurally similar compound. The compounds are often more effective as inducers but are too expensive for use in commercial production. In such cases, constitutive mutants of the organism may be preferable.

An enzyme whose synthesis is inhibited due to **catabolite repression** may be produced by growing the organism in a medium containing a less readily metabolizable form of carbon like starch instead of glucose, or alternatively by feeding the repressive substrate slowly into the fermentation broth such that its concentration remains at a non-inhibitory level. Repression of enzyme production due to product inhibition should also be borne in mind while designing the medium.

5. Fermentation

The fermentation conditions for microbial cultivation are determined by the microorganism being used for enzyme production. Two types of fermentation processes are used. Submerged fermentation, which involves growth of a microorganism as a suspension in a liquid medium, dominates the process of microbial growth for enzyme production in Western countries. In the more traditional **koji** process or solid state fermentation, used in countries of South and South-East Asia, Africa and Central America, microorganisms are growth on solid or semisolid media.

surface-active agent
['sɜːfɪs'æktɪv 'eɪdʒənt]
n. 表面活性剂

inducer [ɪn'djuːsə]
n. 诱导物

catabolite repression
[kə'tæbəlait ri'preʃən]
n. 分解代谢物阻遏

koji ['kəʊdʒɪ]
n. 成曲，制曲，种曲

(1) Submerged fermentation A typical process of enzyme production by submerged microbial fermentation includes the stages of seed preparation, inoculum development and production. The microbial strain is preserved in a form so that its properties are conserved. For this, the organism is either dried by **lyophilization** or on a **sterile inert** carrier, or maintained as suspension in a liquid state at −80℃ in the presence of 20~50 percent glycerol. The first step, therefore, is to revive the culture by planting on agar plants/slants and incubating under appropriate conditions for growth.

lyophilization [laɪˌɒfəlaɪˈzeɪʃən]
n. 冻干法
sterile [ˈsteraɪl]
adj. 无菌的
inert [ɪˈnɜːt]
adj. 惰性的

The microbial culture normally undergoes several stages of inoculum build-up before transfer to the production vessel. The first stage is done in a shake flask, into which cells or spores from agar medium are aseptically inoculated, while the subsequent stages could be in a shake flask or a fermenter under controlled environment. After growth under defined conditions, the culture is transferred into the subsequent level of inoculum or into the production vessel under sterile air pressure or by pumping. The inoculum is usually added to a level of 1%~10% v/v of the culture medium.

After termination of the fermentation, the broth is rapidly cooled from the fermentation temperature to about 5~10 ℃ for maintaining strict **hygienic** measures to control infections during recovery of the enzymes.

hygienic [haɪˈdʒiːnɪk]
adj. 卫生的，保健的

(2) Solid state fermentation The solid fermentation (SSF) process involves the growth of microorganisms on a predominantly insoluble substrate without a free liquid phase. Agricultural raw materials and by-products e.g. **wheat bran**,

wheat bran [hwiːt bræn]
n. 麦麸

rice bran, etc. are frequently used as substrates for SSF. The agricultural products are a rich source of carbon and nitrogen, and contain a variety of other nutrients encapsulated in the biopolymer structure.

Solid state fermentation is done in flasks and petri-dishes in lab scale, while on a larger scale different types of fermenters are used. The simplest SSF system is the **tray** fermenter in which perforated trays covered with a bed of moistened and inoculated growth medium are incubated in an environment of certain temperature and relative humidity. The system can be aerated by blowing air through the medium. Scale up is achieved by increasing the number of trays. The system is simple but also a disadvantage in that the substrate is inefficiently used due to the non-uniform conditions across the fermenter bed.

Microbial enzymes are produced mainly by submerged fermentation under tightly controlled environmental conditions. However, solid-state fermentation (SSF) has also a good potential for the production of enzymes, especially those from filamentous organisms that are particularly suited for surface growth. Some enzymes, particularly those related to **lignocellulose** degradation, are currently being produced by SSF, but other hydrolases, namely amylases, proteases and **phytase** are being produced by SSF as well. SSF compares favorably with submerged fermentation in term of energy requirements, volumetric productivity and product recovery; it represents a good option when production costs should be reduced as is the case of the microbial enrichment of agricultural residues or the production of bulk inexpensive enzymes.

rice bran [raɪs bræn]
n. 米糠

tray [treɪ]
n. 浅盘

lignocellulose [ˌlɪgnəʊˈseljʊləʊs]
n. 木质纤维素
phytase [ˈfaɪteɪz]
n. 植酸酶

6. Regulations during enzyme production

Enzyme products must meet strict specifications with regard to toxicity and other safety aspects. Quality of production of therapeutic enzymes is controlled through pharmaceutical GMP (good manufacturing practice) with strict demands on hygienic standards during all stages of production, process validation and documentation. While production of food and feed enzymes has to comply with Food GMP regulations with a clear focus on **sanitary** processing, production of technical enzyme is not subject to such regulations. For all these enzymes, quality and efficient production are ensured by confirming to ISO9000 standards. As bulk enzyme products undergo a relatively crude purification process, they contain all components of the fermentation broth in small or large quantities. It is therefore necessary to ensure that no toxic products are formed at any stage of the process, e.g. by the metabolism of enzyme generating organism. This is normally controlled during the strain development stage. During fermentation process contamination tests are regularly performed, and contaminated batches discarded. Regular controls of the microbial standard of the finished product are also essential. The residual biomass is normally added to the fields as a **fertilizer** after the process, hence it is essential that it contains no toxic and non-biodegradable chemicals. Use of recombinant microorganisms for enzyme production also puts focus on the prospects of their release into the environment. It need to be ensured that they are safe and do not survive in the general environment. Environmental issues have special focus in case of

sanitary ['sænətri]
adj. 卫生的，清洁的

fertilizer ['fɜːtəlaɪzər]
n. 肥料，化肥

bulk enzymes because of relatively larger production scale, hence reduction in the environment impact from processing is done according to ISO 14000 standards. 　　Due to the **antigenic** nature of the enzyme products and risks of allergy, appropriate formulations like liquid or encapsulated products are essential, so that the exposure of the user to the enzyme is minimized.	antigenic ['æntɪdʒenɪk] 抗原性

参考译文

1　概要

　　在 20 世纪，商品酶的生产规模和产品数量不断增加，以满足不断扩大的市场及其对新型酶的需求。微生物是酶的主要来源。传统酶的生产是采用野生菌株作为生产菌株，然而，通过基因工程技术选育的生产菌株和转基因植物，它们为生产大量酶制剂提供了一个新的选择。

　　一种新的微生物酶的生产，首先要采用合适的菌种选育程序对微生物菌株进行筛选，使微生物产酶的活性达到一个理想值。来源于自然环境的生物体产酶水平通常是低的，这些生物体产酶水平提高后才能用于工业生产。常常采用微生物突变的手段来提高微生物的产酶。另一种提高微生物产酶的策略是采用重组微生物，它能更容易实现微生物生长条件的优化和建立菌种的 GRAS（安全物质）标志。酶的生产之前，采用随机或者位点定向诱变技术，来达到酶的生产中所需活性和稳定性的要求，这已成为一种常见做法。产酶微生物是采用最适培养基在发酵罐内培养。商业酶的生产通常采用固态和液体深层发酵方式。在许多国家都优先使用液体深层发酵方式产酶，因为其在无菌条件下和发酵过程中具有良好的可控性。由微生物产生的酶可以是在细胞内，也可以分泌到细胞外。

　　酶的分离纯化既是酶生产的下游工艺，也是酶生产的后续关键阶段。酶纯化的纯度取决于酶的应用要求。工业批量的酶是粗制的，而特种酶的纯化非常彻底，最终得到单一的产品。传统下游纯化工艺涉及固体原料中酶的沉淀分离、浓缩和分离纯化。酶在细胞内，分离纯化的第一步将细胞或组织破碎，使酶从组织或细胞中释放出来。分离纯化的每个步骤中都可以选择不同的方法。色谱法是纯化高纯度酶的主要技术方法。有些分离方法可以涵盖了几个下游纯化步骤，这些方法可降低产品的成本和减少纯化步骤。酶制剂最终制备成液体或固体产品。不管在哪种情况下，在酶中添加保护剂可以延长保质期。还有一些酶固定化，这些固定化酶具有不溶于水和稳定的特点，可以反复或长时间使用。

2. 酶的来源

酶的生产中首要考虑的是酶源。在大多数情况下，酶所需的活性可来源于几种酶源。然而，传统意义上的酶，其来源更容易受到限制。例如，目前凝乳酶还只能来源于哺乳犊牛的胃，通过微生物方法生产的凝乳酶会使乳酪产品有异味。

微生物是具有吸引力的一种酶源，因为它在相对较短的时间内，采用发酵方法实现大批量生产。但是，不同的微生物中酶的生产水平也是不同的，而且，以上所述酶通常组成和性质也存在不同。人们通常认为密切相关的生物体具有类似属性的酶，而无关的生物体有千差万别的酶系统。用于生产重要的工业酶的生物体，其特征是 GPAS（一般认为安全）的标志，这意味着它们必须无毒、无致病性，一般不产生抗生素。

列入 GRAS 中的微生物包括细菌和真菌，数量少于 50 种。具体是枯草芽孢杆菌、地衣芽孢杆菌，以及其他各种杆菌、乳酸杆菌、链霉菌、酿酒酵母和丝状真菌等。以芽孢杆菌为例，选育的突变株不能形成孢子。由于曲霉培养通常是接种真菌的孢子，产酶的霉菌要具备良好的产孢能力。大部分产量大的酶如水解酶是直接分泌到培养基中，而一些青霉素酰化酶和葡萄糖异构酶是胞内产物。在一些酶的应用上，胞内酶可能没有必要从微生物细胞内分离出来，因为微生物细胞本身可以作为酶源。

优选菌在尽可能短的发酵时间内酶产量高。在工业生产使用的生产菌株通常是通过基因操作来改良的，具有高的生产水平。当今业界普遍的趋势是，将所需酶的基因编码转移至安全性高和表达水平高的微生物生产菌株，采用优化的培养基进行培养。从而避免对单个产酶微生物进行优化。

尽管微生物酶源具有优势，但是动物和植物酶源具有经济优势，这可能是动植物酶源含有丰富的酶，也可能是其可将廉价和再生材料如农业和屠宰废弃物转化为附加值高的酶。

3. 菌种的开发

当生产某种酶的微生物确定后，需要几个步骤，将其转化为一个适合商业用途的菌株。从商业用途角度上来看，野生型微生物产酶量太低。影响酶高水平生产的一个主要的障碍是分解代谢物阻遏，当存在容易代谢的碳源如葡萄糖会抑制一些降解酶的生物合成。另一个问题是，培养基中存在某个诱导物时，降解酶才会大量诱导产生。在工业生产中一些诱导物的价格太贵。

为了达到工业生产的要求，菌株应该生长在价廉的培养基中，同时在短时间内能实现连续且高速产酶。发酵液中次要酶和代谢产物的含量应该是最小的，这将有利于酶的回收。最后，生产工厂中，操作工艺对操作人员必须是安全的，同时从工厂中排放的污水不污染环境。这些目标的实现需要菌种特性和工艺参数的优化组合。

菌种特性的优化，也称为菌种开发，是非常具有吸引力的方法。一般来说，只要解决了菌种问题，很多上述提及的限制性问题将会迎刃而解。例如：

（1）酶的产量及（或）生产力会提高；

（2）污染酶在数量上可以减少或消除；

（3）改变产酶菌的代谢，产生的酶是组成酶；

（4）菌种改良后，可以在分解代谢物阻遏或产物阻遏的条件下产酶。此外，酶的特异性可以被改变，酶的活性和稳定性能够增加，通过菌种开发健康和安全性问题也能得到避免。

由于微生物的生产力和代谢功能都可以通过基因控制，菌株开发通过合适的选育机制来实现基因水平的改变。有三种方法用于菌种改良，包括诱变、杂交和重组 DNA 技术。

4. 微生物的生长需求

酶生产的第一步是微生物的培养。微生物培养基的设计是为了提供微生物生长的基本要素。培养基优化可以使产量和酶制备的成本达到理想值。

在工业生产中，重要的是，生产型微生物在价廉培养基上迅速生长，因为培养基物质的费用占发酵成本的 80%。普遍以碳水化合物为碳源。常用价廉的碳源是糖蜜，它是蔗糖的来源；大麦，玉米，小麦和淀粉水解物等是葡萄糖的来源，乳清是乳糖的来源。氮源的来源既可以是简单的无机盐包括铵盐和硝酸盐，也可以是复合物质如大豆粉、玉米浆、蛋白胨等。适当的无机盐可以提供硫、钙等物质。微量元素由制备培养基的自来水提供，或者在培养基配制中添加。某些微生物的生长需要特定的生长因子如维生素。低浓度的表面活性剂如吐温-80，通过促进酶的释放，来提高胞外酶的产量。

如需要诱导物，通常是诱导底物或者底物类似物。这些化合物往往是更为有效的诱导剂，但是在商业生产中使用过于昂贵。在这种情况下，微生物的突变方法提高酶产量略胜一筹。

受到分解代谢物阻遏的酶，在下列情况下可以解除阻遏，采用不易分解的碳源如淀粉代替葡萄糖，或缓慢流加底物至发酵液中，使底物浓度保持在非抑制水平。产物抑制酶的生物合成，在培养基设计时也应该考虑这个因素。

5. 发酵

微生物的发酵条件取决于酶生产中的微生物。发酵工艺有两种类型，其中液体深层发酵中，微生物在液体培养基中悬浮生长，这种发酵工艺在西方国家占主导地位。传统曲工艺或固态发酵，在南亚、东南亚、非洲和美国中部等地采用，微生物是在固态或半固态基质上生长。

（1）液体深层发酵　深层发酵产酶的典型工艺包括种子制备、种子扩大培养和生产阶段。微生物菌种采用保藏方法保存，以保持其菌种特性。为此，菌种通过冷冻干燥或在无菌惰性载体上干燥，或者在 -80℃下悬浮在 20% ~ 50% 的甘油中。种子制备的第一步是活化，将种子接种到琼脂斜面上，在合适的培养条件下生长。

接种物通常要经过几个阶段进行接种物扩大培养，然后再转移至生产容器。第一阶段是用摇瓶培养，将琼脂培养基中的细胞或孢子通过无菌操作接种。随后的阶段是在可控的条件下进行摇瓶培养或发酵罐发酵。这些接种物制备好后，培养物进一步扩大培养或用无菌空气压/泵到发酵罐中，发酵罐的接种量通常是1%～10%（体积分数）。

发酵结束后，发酵液迅速冷却至 5～10℃，在酶的回收过程保持严格的卫生措施以防止污染。

（2）固态发酵　固态发酵（SSF）工艺，微生物在完全不溶于水和没有游离液相的基质上生长。农业原料和副产物如麦麸、米糠、木薯粉等经常作为固态发酵的基质。农产品富含碳源和氮源，并含有多种营养成分。

实验室规模固态发酵是在摇瓶和培养皿中进行，更大规模则使用不同类型的发酵罐。最简单的固态发酵系统是托托盘式发酵罐，和接种的湿润培养基覆盖在有孔的托盘上，微生物在一定的温度和相对湿度环境中生长。该发酵系统可以透过在培养基吹入空气。增加托盘的个数可以扩大生产规模。这个发酵系统是简单的，但是存在一个劣势，就是由于发酵床中发酵条件不均匀而导致培养基的利用不完全。

微生物酶的生产主要采用能严格控制发酵条件的液态深层发酵。然而，固态发酵产酶也是有很好的潜力，特别是那些来自丝状菌适合表面生长。某些酶，特别是那些木质纤维素降解相关的酶，目前通过固态发酵生产。还有其他水解酶，即蛋白酶和植酸酶也是由固态发酵生产。固态发酵在能源需求、容积生产率和产品回收上，与深层发酵相比毫不逊色。当微生物在农业废弃物上富集或价格低廉酶的批量生产的情况下，要求尽量较少生产成本，那么固态发酵也是一个很好的选择。

6. 产酶规程

酶制剂产品必须符合毒性和其他安全方面的严格规范要求。治疗性酶的生产质量符合药品GMP(良好生产规范)，在生产的各个阶段包括生产、工艺验证和记录对卫生标准严格要求。而食品和饲料酶制剂的生产必须符合食品GMP法规，重点明确的是卫生加工，技术酶的生产不受此类规定。所有的这些酶，ISO 9000标准是保证优质高效的生产。由于大部分酶的生产要经过一个相对粗的纯化过程，这些酶或多或少的含发酵液。因此，有必要确保在生产的任何阶段都没有形成有毒的产物，例如产酶生物体的新陈代谢阶段。这通常在菌种开发阶段进行控制。在发酵过程中定期进行污染测试，并将污染的批次丢弃。成品微生物标准的常规控制也很重要。发酵结束后，菌体残渣通常作为肥料加入到田里。因此重要的是，菌体不含任何有毒和不可生物降解的化学物质。使用重组微生物用于酶的生产，也把重点放在它释放到环境的应用前景。需要确保它们是安全的，在一般环境下无法生存。环境问题特别聚焦在批量酶的方面，因为其有相对较大的生产规模，因此，根据ISO 14000标准可以减少生产过程对环境的影响。

由于酶的抗原性，有过敏的风险，因此，制备合适的制剂如液体制剂或封装产品是必不可少的，从而使用户对酶的曝光最小化。

Further Reading
Relevant Parameters for Enzyme Production

Some relevant aspects to be considered for developing a fermentation process for enzyme production are now analyzed.

Enzyme localization with respect to the producing microorganism is a key aspect in enzyme production. The enzyme can be properly intracellular, **periplasmic** or excreted into the medium during fermentation and this will define the downstream operations for its production. Most enzymes are intracellular but among extracellular enzymes there are many of technological significance; actually, a significant part of the commodity enzymes are extracellular. There are enzymes that are intracellular in one organism and extracellular in another; for instance, **invertase** is mainly intracellular in *Saccharomyces*, while a significant portion of it is excreted in *Candida* and *Streptomyces*; ***β*-galactosidase** is extracellular in molds while intracellular in yeasts. Intracellular enzymes can be made extracellular by genetic engineering and protein engineering techniques.

Specific activity (units of enzyme activity per unit mass of microorganism) is very relevant parameter for enzyme production by fermentation and much effort has been devoted to increase it by both genetic and environmental manipulations. Conventional mutation and selection, genetic engineering, site-directed mutagenesis and directed evolution are powerful genetic tools to obtain high producing microbial strains; in some cases, a substantial portion of the total protein synthesized by the organism corresponds to the enzyme. High specific activity not only reduces the cost of fermentation but also the cost of downstream operations. Significant increase in enzyme specific activity can be obtained by adequate environmental manipulations, mainly through medium design and optimization of relevant operation parameters like temperature, pH, agitation and aeration rates. Enzyme synthesis is subjected to different types of control by the producing strain, so by proper medium design the biological signals that trigger such mechanisms can be put under our control. Specific growth rate of the producing strain is also a relevant parameter for enzyme production by fermentation. Many enzymes are synthesized as growth-associated metabolites so that cell specific growth rate has a direct impact on enzyme specific rate of synthesis.

Genetic stability and safety of the producing microbial strain are also relevant aspects to be considered in enzyme production. This is particularly so in the case of recombinant enzyme proteins because of structural and **segregational** instability of the cloning vector. Depending on the use of the enzyme, the producing strain must be considered safe for its application. For instance, enzymes used in the food industry in USA should have the GRAS (generally recognized as safe) status conferred by the Food and Drug Administration (FDA). To obtain a GRAS status for an organism is costly and time-consuming so that sometimes it is a better option to clone the enzyme structural gene into a GRAS host.

Morphological and **rheological** properties of the producing strain are also relevant for enzyme production, especially for the case of mycelial microorganisms. **Viscosity** increase and **non-Newtonian** rheology may reduce oxygen transfer rates and enzyme synthesis is usually related to one particular growth morphology.

New Words

periplasmic [ˈperɪplæzəm] *adj.* 周质的，胞质的
invertase [ɪnˈvɜːteɪs] *n.* 转化酵素
Candida [ˈkændɪdə] *n.* 念珠菌
β-galactosidase [gəlæktəʊˈsaɪdeɪs] *n.* β- 半乳糖苷酶
specific activity [spɪˈsɪfɪk ækˈtɪvɪti] *n.* 比活性
segregational [segrɪˈgeɪʃənəl] *adj.* 隔离的
rheological [riːəˈlɒdʒɪkəl] *adj.* 流变的
viscosity [vɪˈskɒsətɪ] *n.* 黏度
non-Newtonian rheology [nɒn-njuːˈtəʊniən rɪˈɒlədʒɪ] *n.* 非牛顿流变

参考文献

[1] Jan B van, Beilen, et al. Enzyme technology: an overview. Current Opinion Biotechnology. 2002, 13:338-344.

[2] Andrés Illanes, Lorena Wilson, et al. Problem solving in Enzyme Biocatalysis. John Wiley & Sons, 2013.

[3] Drauz, Karlheinz Gr ger, et al. Enzyme Catalysis in organic Synthesis. Wiley-VCH Verlag GmbH, 2012.

[4] Relx Group. Biofuel speeds ahead with enzyme technology. Oils and Fats International, 2011, 27(2):23-24.

Chapter 3 Immobilized Enzyme Technology 酶的固定化

1. Immobilization techniques

(1) Reversible enzyme immobilization As the name suggests, this method is reversible, and the enzymes can be removed from the support easily by simple reactions or reversal of the conditions by which the immobilization was carried out. The method can be of the following types:

① Adsorption: Adsorption (Figure 5-1) is the oldest and arguably the simplest of all techniques. The first industrial process that involved immobilized enzyme used an **aminoacylase** adsorbed to a DEAE-**Sephadex** for continuous **resolution** of amino acids. The first record of large-scale industrial utilization of immobilized enzyme technology also involved adsorption of glucose isomerase to DEAE-Cellulose in the production of high fructose corn **syrup** by Clinton Corn Products.

aminoacylase [æmɪ'nəʊsiːleɪs]
n. 氨基酰化酶
sephadex ['sefədeks]
n. 葡聚糖凝胶
resolution [ˌrezə'luːʃn]
n. 分辨率
syrup ['sɪrəp]
n. 糖浆，糖汁

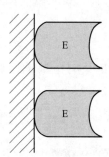

Figure 5-1 Adsorption in immobilized technique

Principle: This involves adhering of the enzyme to the surface of the carrier via several weak non-covalent interactions such as hydrogen bond, Van Der Waal's interactions and **hydrophobic** interaction.

Advantage: This method is cheap, easy to perform and allows easy recovery of the enzyme from

hydrophobic [ˌhaɪdrə'fəʊbɪk]
adj. 疏水的

the carrier, thus allowing re-use of both. The weak interactions involved, hardly cause any distortion of the enzyme retaining maximum enzyme activity.

Disadvantage: Significant enzyme loss cannot be avoided in this technique as the binding forces are weak, reversible and susceptible to physical parameters such as pH and temperature. This may lead to the presence of enzyme in the reaction product, which can contaminate the products and complicate the purification process.

Carriers used: Common carriers used in adsorption are activated charcoal, **alumina**, cellulose, Sephadex, **agarose**, **collagen** and starch. Researchers have come up with new eco-friendly carriers, such as **coconut fibre** with high water retention and **cation** exchange properties, which could significantly reduce costs.

② Disulphide bonding: This technology involves the formation of **disulphide** bonds between the enzyme and the matrix. Though it is a form of covalent bonding (an irreversible enzyme immobilization method), it is classified as a reversible technique because of the ease of reversal of the binding (Figure 5-2).

alumina [əˈluːmɪnə]
n. 氧化铝
agarose [ˈɑːɡərəʊs]
n. 琼脂糖
collagen [ˈkɒlədʒən]
n. 胶原
coconut fibre [ˈkəʊkənʌt ˈfaɪbə]
n. 椰壳纤维
cation [ˈkætaɪən]
n. 阳离子
disulphide [daɪˈsʌlfaɪd]
n. 二硫化物

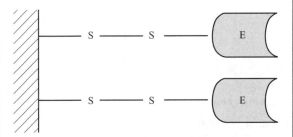

Figure 5-2 Disulphide bonding in immobilized technique

thiol group [ˈθaɪəʊl ɡruːp]
n. 巯基
cysteine [ˈsɪstɪn]
n. 半胱氨酸

Principle: The enzyme immobilization step requires formation of disulphide linkage between the carrier and a free **thiol group**, usually on **cysteine** residues. The disulphide bond is reasonably stable, especially under physiological conditions (at which many enzymes function). The binding can be reversed by the addition of reagents such as **dithiothreitol** (DTT) under mild conditions, or by altering the pH.

dithiothreitol [ˌdɪθɪəʊˈθreɪtəl]
n. 二硫苏糖醇

Advantage: Proper maintenance of pH and temperature can restrict enzyme-leakage, as the disulphide bonds are sufficiently stable. The activity of the thiol group can also be altered with pH.

Disadvantage: In a reaction mixture, the pH and substrate concentration constantly change as the reaction progresses. These may alter the enzyme binding, consequently leading to enzyme loss.

Carriers used: Supports used are generally inert substances like silica, which are chemically activated by agents.

③ Ionic binding: This is a simple reversible mode of immobilization of proteins, which involves ionic interaction between the enzyme and the support (Figure 5-3).

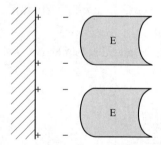

Figure 5-3　Ionic binding in immobilized technique

Principle: The support used is generally charged, such that the protein to be bound has an opposite charge. The enzyme is therefore bound to the support via ionic interactions. It can be easily reversed by altering the pH or "salting out" of the enzyme.

Advantage: It is very easy, inexpensive and requires simple inputs for reversal of the binding.

To maintain an optimum pH during the reaction tenure, easy manipulation of the acidity or **alkalinity** in the reaction mixture can be performed, as the matrix which immobilizes the enzyme is stably charged.

Disadvantage: The presence of the charged support causes several problems like enzyme structure distortion and alterations in enzyme kinetics. High charge has the potential to disrupt the enzyme catalysis. As a result, maximum yield is hindered.

④ Affinity binding: This technique is based on the **antigen-antibody** interaction (Figure 5-4).

alkalinity [ˌælkəˈlɪnəti]
n. 碱性

antigen-antibody [ˈæntiːdʒenˈæntɪbɒdɪ]
n. 抗原－抗体

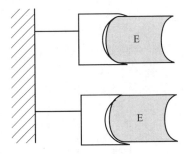

Figure 5-4 Affinity binding in immobilized technique

Principle: This technique is based on high affinity interaction between biomolecules. The carrier matrix is synthesized specifically for a single type of enzyme and contains antibodies against specific **epitopes** on the antigen (enzyme).

Advantage: The reaction is highly specific and no contaminants are present on the carrier. If the antibody on the support is highly specific for the enzyme, the step of enzyme purification can be bypassed. Enzymes from an impure solution can also specifically attach to the matrix. Maximum activity of the enzyme is also ensured if the antibody is targeted at an epitope away from the activity site.

epitope [ˈepɪtəʊp]
n. 抗原位点

Disadvantage: The method involves use of specific antibodies, which are generally very expensive.

(2) Irreversible enzyme immobilization
Irreversible immobilization involves strong chemical bonds and particularly serves to maintain reasonable stability of the enzymes over a long period of time. Most industries use enzymes immobilized by these methods.

① Covalent binding: **Covalent bonds** are highly stable and hence, covalent binding ensures that the enzyme is strongly bound to the support (Figure 5-5).

covalent bond
[kɔʊˌveɪ.lənt 'bɑːnd]
n. 共价键

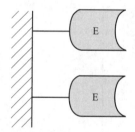

Figure 5-5　Covalent binding in immobilized technique

Principle: It involves the formation of a covalent bond between the support and the side chains of the amino acids of the enzyme, most commonly **lysine** (å-amino group), **cysteine** (thiol group), etc.

Advantage: The binding involves covalent interactions and is strong. Hence, leaking of the enzyme into the reaction mixture is totally prevented. This prevents mixing of the enzyme with the product, thereby reducing contamination and the cost of purification. The covalent binding also stabilizes the enzyme in specific protein orientations, and may promote higher specific activity.

lysine ['laɪsiːn]
n. 赖氨酸
cysteine ['sɪstɪn]
n. 半胱氨酸

Disadvantage: The covalent bond formed between the support and enzyme may involve the amino acids of the active site of the enzyme, which may lead to significant loss in activity. Since the method is irreversible, the support cannot be recycled, as the enzymatic activity declines. The support along with the bound enzyme has to be discarded.

Carriers used: The supports used are generally stable and easily available and are activated by the appropriate reagents. The common supports used are **cyanogen bromide** (CNBr)-activated Sephadex or CNBr-activated **Sepharose**. Other common carriers include activated forms of **dextran**, cellulose, agarose, etc. Artificial matrices include **Polyvinyl chloride**, ion exchange **resins** and porous glass.

② Crosslinking: It is an irreversible method of enzyme immobilization (Figure 5-6). It is different from other techniques in the sense that it does not require a support for the immobilization. Cross linking agents such as **glutaraldehyde** (which react with the amino group on the protein) were used. Hence, unlike the other systems of enzyme

cyanogen [saɪˈænədʒɪn]
n. 氰，氰化物
bromide [ˈbrəʊmaɪd]
n. 溴化物
sepharose [ˈsefərəʊs]
n. 琼脂糖凝胶
dextran [ˈdekstrən]
n. 葡聚糖
polyvinyl chloride
[ˌpɒliˈvaɪnɪl ˈklɔːraɪd]
n. 聚氯乙烯
resin [ˈrezɪn]
n. 树脂，合成树脂
glutaraldehyde
[ˌɡluːtəˈrældɪhaɪd]
n. 戊二醛

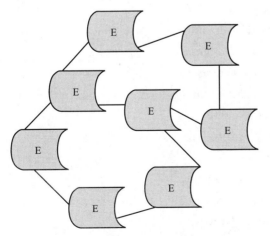

Figure 5-6　Crosslinking in immobilized technique

immobilization, the immobilized enzyme is not bound to any matrix, but is present in the reaction mixture, albeit in an immobilized form.

Principle: In this method, glutaraldehyde is used to crystallize the enzyme. Hence, enzyme crystals are obtained with the help of the cross-linking agent. The Cross Linking Enzyme Crystals (CLEC), upon addition to the reaction mixture, catalyzes the reaction with reasonably high efficiency.

Advantage: CLEC are very stable and are not easily denatured by heat or organic solvents. They are moderately resistant to **proteolysis**. They have a manageable size and stability in operating the bioreactor and can be recycled.

Disadvantage: Highly purified enzyme produced through a standard protocol of crystallization is required for the preparation of CLEC. These requirements involve a lot of time and expenses. Diffusion of substrate and product is limited with increase in size of the aggregate.

Cross Linking Enzyme Aggregates (CLEA) is an improved version of CLEC production and aims at nullifying the disadvantages of CLEC. While CLEC requires the formation of crystals, CLEA could work in aqueous solutions. Addition of salts, organic solvents or non-ionic polymers results in the formation of enzyme aggregates which retain their catalytic properties. These aggregates are called Cross Linked Enzyme Aggregates (CLEA).

Crosslinking agents used: For most enzymes, the crosslinking agent used is glutaraldehyde, which is cheap, stable and easily available. However, glutaraldehyde partially or totally inactivates some

proteolysis [ˌprəʊtɪˈɒləsɪs] *n.* 蛋白水解，蛋白质水解

enzymes. For such biomolecules, alternative cross-linking agents such as dextran **polysaccharide, bis-isocyanate**, et al, and functionally inert proteins, such as **bovine serum albumin** (BSA) should be preferred.

③ **Entrapment:** Entrapment is another prominent technique of irreversible enzyme immobilization, where the enzyme is immobilized by entrapping it within a support matrix or within fibres (Figure 5-7).

polysaccharide [pɒlɪˈsækəraɪd]
n. 多糖，聚糖
bis-isocyanate [bɪsaɪsəʊˈsaɪəneɪt]
n. 二异氰酸酯
bovine serum albumin [ˈbəʊˌvaɪn ˈsɪərəm æl'bjumin]
n. 牛血清白蛋白
entrapment [ɪnˈtræpmənt]
n. 包埋

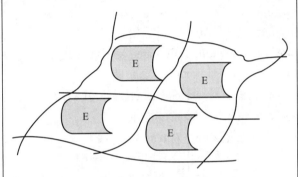

Figure 5-7　Entrapment in immobilized technique

Principle: Enzymes, being large macromolecules, tend to be larger than the substrates or products. Thus, the enzyme is immobilized within a matrix of appropriate pore size to allow only the substrates and products of a diameter smaller than the matrix pore size to diffuse in and out of the mesh respectively. The enzyme size-to-pore size of support is a deciding factor in selecting the support. Smaller the pores, lesser the enzyme entrapped, while larger the pores, more the leaking of the enzyme. Hence, accurate pore size selection of the support is crucial.

Advantage: The method is fast, cheap and easily carried out under mild or physiological conditions. As the enzyme remains confined within a matrix, it is protected from contamination by microbes, proteases or other enzymes.

Disadvantage: The meshwork of the matrix cannot support a huge volume of enzyme molecules and can lead to enzyme inactivation. Hence, the process can be costly at times. The rate of diffusion of the substrate and product dictate the reaction rate. This is because unless the substrate molecules diffuse into the mesh, the reaction will not be initiated and according to **Le Chatelier's principle**, the reaction rate does not reach a peak unless the products **sieve** out.

Matrix used: Common polymers used for enzyme entrapment include **alginate**, **carrageenan**, collagen, **polyacrylamide**, **gelatin**, and so on.

Encapsulation can be regarded as a special type of entrapment where the enzyme is immobilized by entrapping it in a spherical semi-permeable membrane (Figure 5-8).

Le Chatelier's principle
勒夏特列原理
sieve [sɪv]
v. 筛，滤
alginate ['ældʒɪneɪt]
n. 海藻酸钠
carrageenan [kærə'giːnən]
n. 卡拉胶
polyacrylamide [pɒlə'krɪləmaɪd]
n. 聚丙烯酰胺
gelatin ['dʒelətɪn]
n. 明胶
encapsulation [inˌkæpsju'leiʃən]
n. 微囊

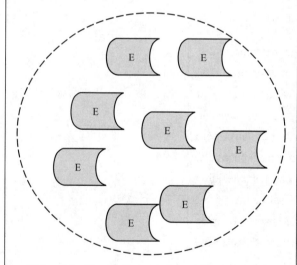

Figure 5-8　Encapsulation in immobilized technique

2. Benefits and characteristics

There are several reasons to use immobilized enzymes. In addition to the convenient handling of enzyme preparations, the two main targeted benefits are: ① easy separation of enzyme from the product; and ② reuse of the enzyme. Easy separation of the enzyme from the product simplifies enzyme applications and permits reliable and efficient reaction technology. Enzyme reuse provides a number of cost advantages, which are often an essential **prerequisite** for establishing an economically viable enzyme-catalyzed process.

prerequisite [ˌpriːˈrekwəzɪt] *n.* 先决条件，前提

The properties of immobilized enzyme preparations are governed by the properties of both the enzyme and the carrier material. The interaction between the two provides an immobilized enzyme with specific chemical, biochemical, mechanical and kinetic properties.

As far as manufacturing costs are concerned, the yield of immobilized enzyme activity is determined by the immobilization method in relation to the amount of soluble enzyme used. Under process conditions, the resulting activity can be further reduced by mass-transfer effects. That is, the yield of enzyme activity following immobilization does not only depend on losses caused by the binding procedure but can be further reduced as a result of the diminished availability of enzyme molecules within pores or by slowly diffusing substrate molecules. Such limitations lead to lowered efficiency. However, improved stability under working conditions can compensate for such drawbacks, resulting in an overall benefit.

参考译文

1. 固定化技术

（1）可逆的固定化酶方法　顾名思义，该方法是可逆的，并且固定化酶可以通过简单的反应或固定化的逆转条件将酶从载体上解除。该方法有以下类型。

① 吸附：吸附（图 5-1）是最古老和最简单的技术。第一个用于工业生产的

就是将氨酰化酶吸附到 DEAE- 葡聚糖凝胶，制备的固定化酶可连续分辨氨基酸。固定化酶技术大规模工业应用首次记载是，克林顿玉米制品公司将葡萄糖异构酶吸附到 DEAE- 纤维素生产高果糖玉米浆。

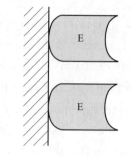

吸附原理：通过非共价键弱相互作用如氢键、范德华相互作用和疏水相互作用将酶吸附到载体的表面。

图 5-1 固定化技术中的吸附法

优点：方法简单，操作简便，酶易于回收和可重新使用。酶与载体间弱的相互作用很难导致酶变形，保证了酶的最高活性。

缺点：酶不可避免会损失，因为作用力弱且可逆，且容易受外界因素（pH 和温度）影响。这可能导致反应产物中含有酶，它会污染产品和导致纯化过程复杂化。

载体：吸附中常用的载体是活性炭、氧化铝、纤维素、葡聚糖、琼脂糖、胶原和淀粉。研究人员已经提出生态友好的新载体，如具有高持水性和阳离子交换性能的椰子纤维，这可明显降低成本。

② 二硫键：这个固定化方法是酶与载体之间形成二硫键（图 5-2）。虽然是共价结合（不可逆的酶固定化方法），但因这种结合易可逆而把它归于可逆的固定化方法。

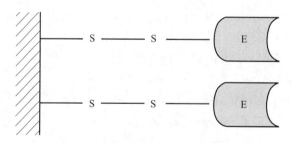

图 5-2 固定化技术中的二硫键结合法

原理：这种固定化方法需在载体与半胱氨酸残基的游离巯基之间形成二硫键。特别是在生理条件下（许多酶的作用下），二硫键是相当稳定的。载体与酶的结合可在温和的条件下通过添加二硫苏糖醇（DTT）试剂或改变 pH 发生逆转。

优点：维持适当的 pH 和温度可以防止酶的泄漏，二硫键是相当稳定的。巯基活性可通过 pH 改变。

缺点：在反应中，pH 和底物浓度不断发生改变。这可能会改变酶的结合，而导致酶的损失。

载体：一般是惰性物质如二氧化硅，用化学试剂激活。

③ 离子结合：这是简单可逆的蛋白质固定化方法，酶与载体之间是离子相互作用（图5-3）。

原理：使用的载体通常是电荷，与该蛋白质结合的载体具有相反电荷。因此，酶通过离子相互作用与载体结合。改变pH或酶的盐析很容易发生可逆反应。

优点：操作容易，价格低廉，可逆反应操作简单。在反应中调节酸度或碱度易于调节pH，载体固定化酶带电稳定。

缺点：电荷载体会导致一些问题，如酶结构的变形和酶动力学的改变。高电荷有可能会破坏酶的催化作用。其结果是阻碍最大产量的产生。

④ 亲和结合：这种技术是基于抗原-抗体相互作用（图5-4）。

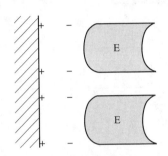 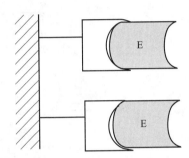

图5-3　固定化技术中的离子结合法　　　图5-4　固定化技术中的亲和结合法

原理：该技术是基于生物分子之间的高度亲和力作用。载体基质是根据酶的特异性合成，并含与抗原（酶）相对应的抗体。

优点：该反应是高度特异性，载体上没有污染物。如果载体上的抗体对酶有高度特异性，那么酶的纯化步骤可以省略。含有杂质的酶液可以特异性吸附到基质上。如果抗体能够从活性位点脱离，那么可以保证酶的活性。

缺点：该方法所使用的是特异性抗体，价格通常是非常昂贵的。

(2) 不可逆的固定化酶方法

① 共价结合：共价键非常稳定，因此，酶非常强的结合到载体上（图5-5）。

原理：载体和氨基酸侧链之间形成共价键，最常见的侧链是赖氨酸（ε-氨基）、半胱氨酸（巯基）等。

优点：共价结合力强，完全防止酶的泄露。防止了酶与产品混合，减少污染和纯化的费用。共价键稳定酶蛋白的排列方向，可提升酶的比活性。

缺点：载体与酶形成的共价键可能会涉及酶的活性位点，这可能导致酶活性大的损失。由于该方法不可逆，载体不能回收。

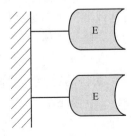

图5-5　固定化技术中的共价键结合法

载体：使用的载体通常是稳定的，容易获得并且由适当的试剂激活。常见的载体是溴化氰（CNBr）活化的葡聚糖或溴化氰活化的琼脂糖凝胶。其他常见的载体包括活性形式的葡聚糖、纤维素、琼脂糖等。人工合成载体包括氯乙烯、离子交换树脂和多孔玻璃。

② 交联：这是不可逆的固定化方法（图5-6）。与其他技术不同的是，它的固定化不需要载体。采用交联剂如戊二醛（与蛋白质氨基基团反应）。因此，不像其他的酶固定化系统，固定化酶不需要结合到任何载体，是在反应液中呈现固定化形式。

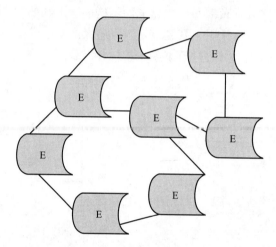

图5-6　固定化技术中的交联法

原理：在该方法中，戊二醛用于酶的结晶。因此，采用交联剂得到酶晶体。交联酶晶体（CLEC）作为催化剂，反应具有相当高的催化效率。

优点：CLEC非常稳定，不容易通过热或有机溶剂变性。中度抗蛋白水解。在生物反应器中，交联酶晶体具有可控的大小和稳定性，也可以回收。

缺点：通过标准结晶方法产生高纯度酶是CLEC的制备要求。这些要求涉及了大量时间和费用。增加晶体大小会限制底物和产品的扩散。

交联酶聚集体（CLEA）是CLEC生产的改进版本。旨在消除CLEC的缺点。CLEC的制备需要形成晶体，而CLEA可以在水溶液中形成。加入盐、有机溶剂或非离子型聚合物，形成酶的聚集体，保留了酶的催化特性，这些聚集体称为交联酶聚集体。

交联剂：对于大多数酶，使用的交联剂是戊二醛，具有价廉的、稳定且容易获得的特点。然而，戊二醛会使酶的活性部分或完全丧失。可采用葡聚糖、二异氰酸酯等来替代戊二醛，功能惰性蛋白如牛血清白蛋白（BSA）是首选。

③ 包埋法：包埋法是不可逆酶固定化方法中突出的技术方法。酶是通过包埋在载体基质或纤维中进行固定化（图5-7）。

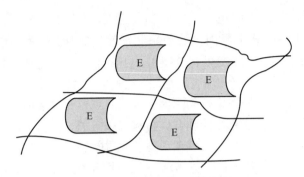

图 5-7　固定化技术中的包埋法

原理：生物大分子酶的尺寸往往比底物或产物要大。因此，将酶固定在适当孔径的载体中，小于孔径大小的底物和产物可通过网格扩散。载体孔径的大小是载体选择的决定性因素。当孔径变小，包埋的酶变少；而孔径变大，更多的酶会泄露。因此，载体孔径的精确选择是至关重要的。

优点：该方法快速、价廉且易于在温和或生理条件下进行。由于酶包裹在载体中，它不受微生物、蛋白酶和其他酶的污染。

缺点：载体的网格结构不能支撑体积巨大的酶，这可导致酶的失活。因此，该固定化过程有时是昂贵的。底物和产物的扩散速度决定反应速度。这是因为，基于勒夏特列原理即平衡移动原理，只有底物进入网格中，要不反应无法启动，只有产物从网格结构中移除反应速率才会达到最高值。

载体：用于酶包埋的共聚物包括海藻酸钠、卡拉胶、明胶、胶原、聚丙烯酰胺、硅橡胶和聚氨酯。

胶囊包埋法是包埋法中的特殊类型，酶是通过球形半透性膜来固定化（图 5-8）。

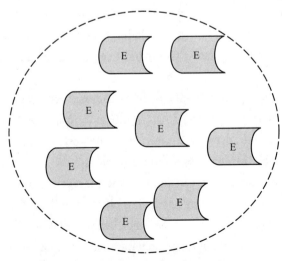

图 5-8　固定化技术中的胶囊包埋法

2. 固定化酶的益处和特点

使用固定化酶有几个原因。除了酶制剂使用方便外，两个主要的优势是：① 产物易于与酶分离；② 酶可重复使用。酶易于与产品分离，简化了固定化酶的应用，并允许有可靠和有效的反应技术。酶的重复使用提供了很多成本优势，这往往是用于建立一种经济可行的酶催化工艺的重要先决条件。

固定化酶制剂的性能是由酶和载体材料来决定的，两者之间的相互作用形成了具有特定化学、生物化学、机械和动力学性质的酶。

就生产成本而言，固定化酶的活力是由固定化方法来决定，该方法与游离酶的使用量有关。在固定化工艺条件下，酶的活力受传质影响会进一步降低，即酶的活力不仅取决于固定化过程造成的损失，还取决于孔隙中或底物分子扩散降低，有效催化的酶分子减少，而导致酶活力进一步降低。这些限制因素会降低酶的催化效率。但是，酶稳定性的提高可以弥补这些缺点，总体是有益的。

Further Reading
Economics of Immobilized Enzymes

The world market for industrial enzymes exceeded $US 3.3 billion in 2010 and is projected to grow to around $4.5 to $5 billion by 2015 at a compounded annual growth rate of 7%~9%. This market is dominated by products containing non-immobilized enzymes, predominantly hydrolase (e.g. amylase, proteases, cellulases and lipases). These products are either liquid concentrates, or enzymes **granules** that release the soluble enzyme upon dissolution. Enzymes for use in non-industrial markets, primarily for pharmaceutical, diagnostic and research applications, accounted for around $2.4 billion in 2010. Sale of enzymes for bio-catalysis, many of which are used in immobilized form, were valued at $160 million in 2010 and projected to increase to $230 million by 2015 (not including captive use by companies producing their own immobilized enzymes). The market segments for enzymes is depicted in Figure 5-9.

Current sales of immobilized enzymes themselves amount to only a small fraction of the total enzyme market, down significantly from 1990, when accounted for nearly 20% of all industrial enzyme sales. Why the apparent disconnect between the thriving scientific field of enzyme immobilization and the very modest market share of immobilized enzymes relative to the enzyme market at large? From the enzyme producer's perspective, the cost of producing an immobilized form of enzyme must enable a new application, or offer some other benefit relative to the soluble form of the enzyme. The fact that immobilized enzymes can often be reused does not directly benefit the enzyme producer, but often provides incentive for customers to purchase an

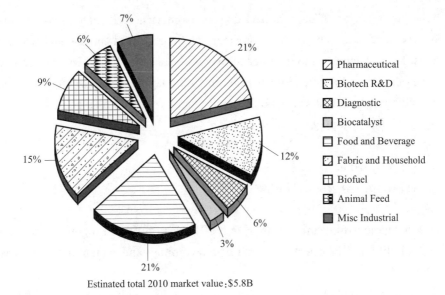

Figure 5-9 A breakdown of the world enzyme market in 2010 by sector

immobilized enzyme product. In some cases, an enzyme producer may attempt to gain some value beyond enzyme sales alone, for example by securing intellectual property on the processes and products derived from the use of immobilized enzyme. In other instances, a customer in need of an immobilized enzyme will purchase a soluble form of the enzyme, and prepare an immobilized form of the enzyme themselves, or outsource this task to companies that specialize in immobilization. Established chemical companies increasingly produce and immobilize their own enzymes in parallel with process development. Overall, product opportunities need to be chosen carefully to ensure that an immobilized enzyme is really needed and offers reasonable return.

It is often incorrectly assumed that industrial enzymes are very expensive and that these costs drive the need for an immobilized, recyclable form of biocatalyst. In reality the cost of most industrial enzymes is in the $50 to $500 per kg enzyme protein range, and they are often only a minor component in overall process economics. For example, the total cost of enzymes for starch-derived ethanol is around 1 cent per liter. In these instances, the additional expenses associated with enzyme immobilization are not worth the return that might be gained from enzyme recycle. The cost contribution from an immobilized enzyme is dependent on the number of times the enzyme is reused, an indirect measure of total productivity on a kg product per kg biocatalyst basis. This amount varies between a few hundred $ per kg for specialty chemicals, down to a few cents per kg for bulk chemicals, and is often in the range of $0.1 to $10 per kg.

The early success of immobilized glucose **isomerase** (IGI) holds several lessons for the design of future immobilized biocatalysts, both isolated enzyme and whole cell products. Many factors came together to drive the success of IGI. This suggests that the development and application of immobilized enzyme products requires a good understanding of both technical and economic factors, as well as a good sense of the larger market forces at play.

New Words

granule ['grænjuːl] *n.* 小颗粒

isomerase [aɪ'sɒməˌreɪs] *n.* 异构酶

参考文献

[1] S Krishnanoorthi, Aditya Banerjee, et al. Immobilized Enzyme Technology: Potentiality and Prospects. Journal of enzymology and metabolism, 2015,1(1): 104-115.

[2] DiCosimo R, McAuliffe J, et al. Industrial use of immobilized enzymes. Chem Soc Rev, 2013, 42(15):6437-6474.

Unit 6　Production of Dosage Forms　药物制剂

Chapter 1　Introduction to Drugs　药物简介

1. What is a drug 　　A drug may be defined as an agent intended for use in the **diagnosis**. **mitigation**, treatment, cure, or prevention of disease in man or in other animals. A drug is any natural or synthetic substance that alters the physiological state of a living organism. 　　Although drugs are intended to have a selection action, this is rarely achieved. There is always a risk of adverse effects associated with the use of any drug. No drug is without side effects, although the severity and frequency of these will vary from drug to drug and from person to person. Those who are more prone to the adverse effects of drugs include: 　　(1) Pregnant women, who must be careful about taking drugs as certain drugs cause fetal **malformations**. 　　(2) Breast-feeding women, who must also be careful about which drugs they take, as many drugs can be passed on in the breast milk and consumed by the developing infant. 　　(3) Patients with liver or kidney disease. These illnesses will result in decreased metabolism and excretion of the drug and will produce the side effects of an increased dose of the same drug. 　　(4) The elderly, who tend to take a large number of drugs, greatly increasing the risk of drug interactions and the associated side effects. In	**New Words and Expressions** diagnosis [ˌdaɪəgˈnəʊsɪs] *n.* 诊断 mitigation [mɪtɪˈgeɪʃən] *n.* 减轻；缓和；平静 malformation [ˈmælfɔrˈmeʃən] *n.* 畸形；变形

addition, elderly patients have a reduced **renal** clearance, and a nervous system that is more sensitive to drugs. The dose of drug initially given is usually 50% of the adult dose, and certain drugs are contraindicated.

  A drug causes a change of physiological function by interacting with the organism at the chemical level.

  Certain drugs work by means of their physicochemical properties and are said to have a non-specific mechanism of action. For this reason, these drugs must be given in much higher doses than the more specific drugs.

  Most drugs produce their effects by targeting specific cellular macromolecules. This may involve modification of DNA or RNA function, inhibition of transport system or enzymes or, more commonly, action on receptors.

  One of the most astounding qualities of drugs is the diversity of their actions and effects on the body. Some drugs selectively stimulate the **cardiac** muscle, the smooth muscles, or the **skeletal** muscles; other drugs have the opposite effect. **Mydriatic** drugs dilate the pupil of the eye; **miotic** constrict or diminish **papillary** size. Drugs can render blood more **coagulable** or less coagulable; they can increase the **hemoglobin** content of the **erythrocytes** or expend blood volume.

  Drugs termed **emetics** induce **vomiting**, whereas antiemetic drugs have the opposite effect. **Diuretic** drugs increase the flow of urine, **sudorific** drugs promote sweating, expectorant drugs increase **respiratory** tract fluid, and **cathartics** or **laxatives** promote the evacuation of the bowel. Other

renal ['rinl]
adj. 肾脏的，肾的

cardiac ['kɑːdɪæk]
adj. 心脏的；心脏病的
skeletal ['skɛlətl]
adj. 骨骼的
mydriatic [ˌmɪdrɪ'ætɪk]
adj. 瞳孔放大的
miotic [maɪ'ɑtɪk]
adj. 瞳孔缩小的
papillary [pə'pɪlərɪ]
adj. 乳突的；乳头状突起的
coagulable [kəʊ'ægjʊləbl]
adj. 可凝结的
hemoglobin [ˌhiːməʊ'gləʊbɪn]
n. 血红蛋白；血红素
erythrocyte [ɪ'rɪθrəʊsaɪt]
n. 红细胞
emetic [ɪ'metɪk]
n. 催吐剂
vomiting ['vɑmɪt]
v. 呕吐
diuretic [ˌdaɪju'rɛtɪk]
n. 利尿剂
sudorific [ˌsjuːdə'rɪfɪk]
adj. 发汗的，促使发汗的
respiratory ['respərətɔːri]
adj. 呼吸的
cathartic [kə'θɑrtɪk]
n. 泻药；通便药
laxative ['læksətɪv]
n. 泻药；缓泻药

drugs may decrease the flow of urine, diminish body **secretions**, or induce **constipation**.

Drugs can be employed to reduce headache, pain, fever, thyroid activity, sneezing, rhinitis, insomnia, gastric acidity, motion sickness, and mental depression. Drugs can elevate the mood, the blood pressure, or the activity of the endocrine glands. Drugs can combat infectious disease, destroy intestinal worms, or act as antidotes against the poisoning effects of other drugs. Antineoplastic drugs provide one means of attacking the cancerous process; radioactive pharmaceuticals provide another.

Drugs may be used to diagnose diabetes, liver malfunction, tuberculosis, or pregnancy, or they may be employed to replenish a body deficient in antibodies, vitamins, hormones, electrolytes, protein, enzymes, or blood. Drugs may be used to prevent measles, poliomyelitis, or pregnancy, or to assist the maintenance of pregnancy or to extend life itself.

Certainly the vast array of effective medicinal agents available today represents one of man's greatest scientific accomplishments. It would be frightening to conceive of our civilization devoid of these remarkable and beneficial agents. Through their use, many of the diseases which have plagued mankind throughout history, as **smallpox** and **poliomyelitis**, are now virtually extinct. Illnesses such as **diabetes**, **hypertension**, and mental depression are now effectively controlled with modern drugs. Today's surgical procedures would be virtually impossible without the benefit of general **anesthetics**, **analgesics**, antibiotics, blood transfusions, and intravenous fluids and nutrients.

secretions [sɪ'kriʃən]
n. 分泌物
constipation [ˌkɑnstə'peʃən]
n. 便秘

smallpox ['smɔl'pɑks]
n. 天花
poliomyelitis [ˌpolɪoˌmaɪə'laɪtɪs]
n. 小儿麻痹症
diabetes [ˌdaɪə'bitɪz]
n. 糖尿病
hypertension [ˌhaɪpə'tɛnʃən]
n. 高血压
anesthetics [ænɪs'θitiks]
n. 麻醉药；麻醉剂
analgesics [ˌænəl'dʒiːzɪks]
n. 镇痛药；止痛剂

2. Drug **bioavailability**

The physicochemical properties of a drug govern its absorptive potential, but the properties of the dosage form (which partly depend on its design and manufacture) can largely determine drug bioavailability. Differences in bioavailability among formulations of a given drug can have clinical significance. Thus, the concept of equivalence among drug products is important in making clinical decisions. Chemical equivalence refers to drug products that contain the same compound in the same amount and that meet current official standards; however, inactive ingredients in drug products may differ. **Bioequivalence** refers to chemical equivalents that, when administered to the same person in the same dosage regimen, result in equivalent concentrations of drug in blood and tissues. Therapeutic equivalence refers to drug products that, when administered to the same person in the same dosage regimen, provide essentially the same therapeutic effect or toxicity. Bioequivalent products are expected to be therapeutically equivalent.

Sometimes therapeutic equivalence may be achieved despite differences in bioavailability. For example, the therapeutic index (ratio of the **maximum tolerated dose** to the minimum effective dose) of penicillin is so wide that moderate blood concentration differences due to bioavailability differences in penicillin products may not affect therapeutic efficacy or safety. In contrast, bioavailability differences are important for a drug with a relatively narrow therapeutic **index**.

bioavailability
['baiəuə,veilə'biləti]
n. 生物利用度

bioequivalence
[,baiəui'kwivələns]
n. 生物等效性

maximum tolerated dose
['mæksɪməm 'tɒləreɪt dəʊs]
n. 最大耐受剂量

index ['ɪndɛks]
n. 指标；指数

The physiologic characteristics and **comorbidities** of the patient also affect bioavailability. Absorption rate is important because even when a drug is absorbed completely, it may be absorbed too slowly to produce a therapeutic blood level quickly enough or so rapidly that toxicity results from high drug concentrations after each dose.

When a drug rapidly dissolves and readily crosses membranes, absorption tends to be complete, but absorption of orally administered drugs is not always complete. Before reaching the **vena cava**, a drug must move down the GI tract and pass through the **gut** wall and liver, common sites of drug metabolism; thus, a drug may be metabolized (first-pass metabolism) before it can be measured in the systemic circulation. Many drugs have low oral bioavailability because of extensive first-pass metabolism. For such drugs (eg, **isoproterenol**, **norepinephrine**, **testosterone**), extraction in these tissues is so extensive that bioavailability is virtually zero. For drugs with an active metabolite, the therapeutic consequence of first-pass metabolism depends on the contributions of the drug and the metabolite to the desired and undesired effects.

Low bioavailability is most common with oral dosage forms of poorly water-soluble, slowly absorbed drugs. More factors can affect bioavailability when absorption is slow or incomplete than when it is rapid and complete, so slow or incomplete absorption often leads to variable therapeutic responses.

Insufficient time in the GI tract is a common cause of low bioavailability. **Ingested** drug is exposed to the entire GI tract for no more than 1 to

comorbidity [kemɔr'bidəti]
n. 并存病

vena cava ['viːnə 'keivə]
n. 腔静脉
gut [gʌt]
n. 内脏；肠子

isoproterenol ['aisəu,prəutə'riːnɔl]
n. 异丙肾上腺素
norepinephrine [ˌnɔrɛpɪ'nɛfrin]
n. 去甲肾上腺素
testosterone [te'stɒstərəun]
n. 睾酮，睾丸素

ingest [ɪn'dʒɛst]
vt. 摄取；咽下

2 days and to the small **intestine** for only 2 to 4 h. If the drug does not dissolve readily or cannot **penetrate** the **epithelial** membrane (eg, if it is highly ionized and polar), time at the absorption site may be insufficient. In such cases, bioavailability tends to be highly variable as well as low. Age, sex, activity, genetic **phenotype**, stress, disease, or previous GI surgery can affect drug bioavailability. Reactions that compete with absorption can reduce bioavailability. They include complex formation (eg, between **tetracycline** and **polyvalent** metal ions), hydrolysis by gastric acid or digestive enzymes (eg, penicillin and **chloramphenicol palmitate** hydrolysis), conjugation in the gut wall (eg, sulfoconjugation of isoproterenol), adsorption to other drugs (eg, **digoxin** and **cholestyramine**), and metabolism by **luminal** microflora.	intestine [ɪnˈtɛstɪn] n. 肠 penetrate [ˈpɛnətret] vt. 渗透；穿透 epithelial [ˌɛpɪˈθilɪəl] adj. 上皮的；皮膜的 phenotype [ˈfinətaɪp] n. 表型 tetracycline [ˌtɛtrəˈsaɪkliːn] n. 四环素 polyvalent [ˌpɒlɪˈveɪlənt] adj. 多价的 chloramphenicol palmitate [ˌklɔːræmˈfenɪkɒl ˈpælmɪteɪt] n. 氯霉素棕榈酸酯 digoxin [dɪdʒˈɔksin] n. 地高辛 cholestyramine [ˌkɒlɪˈstɪrəˌmin] n. 消胆胺 luminal [ˈljuːminəl] adj. 腔的；细胞腔的

参考译文

1. 什么是药物

药物可以定义为用以预防、诊断、缓和、治疗人体与动物体疾病的物质。药物是能改变生物机体生理状态的天然或合成物质。

尽管人们期望药物有选择性作用，但这一点很难达到。任何药物在使用时都存在毒副作用。没有副作用的药物是不存在的，不同的只是在药物与药物之间以及人与人之间毒副作用的严重程度和发生概率不一样。容易受毒副作用伤害的有以下几类人员。

(1) 孕妇　应慎重服用药物，因某些药物会导致婴儿畸形。

(2) 哺乳期妇女　应慎重服用药物，因为很多药物会传递到乳汁，从而进入婴儿体内。

(3) 肝、肾病患者　这些疾病会使药物的代谢和排泄减慢（导致体内药物浓度增加），从而产生相当于药物过量而带来的副作用。

（4）老年人　这类人群通常吃很多药，极大增加了药物之间相互作用的可能性，从而产生相关副作用。另外，老年患者肾清除能力下降，神经系统对药物敏感度增加，药物的初始剂量通常只有成年人的一半，有些药物根本不能给老年人服用。

药物通过与生物体在化学水平上的相互作用改变其生理功能。某些药物借助它们的物理化学性质起作用，即所谓的非特异性功能药。由于没有特定的作用机制，这类药的给药剂量要比特异性功能药大很多。大多数药物通过作用于细胞上的大分子团起作用。这些作用包括修改 DNA、RNA 的功能，抑制（细胞）传输系统或酶系统，或者更普遍的情况是结合到受体上产生作用。

药物最令人惊讶的属性之一是它们的多样性和对身体的影响。一些药物选择性地刺激心肌、平滑肌或者骨骼肌；其他药物产生相反的效果。瞳孔放大的药物扩张眼睛的瞳孔，缩瞳剂收缩或减少视乳头大小。药物可以使血液更易凝结或降低可凝性；它们可以增加红细胞的血红蛋白含量或消耗血容量。

药物可以用来减少头痛、疼痛、发热、甲状腺活动、打喷嚏、鼻炎、失眠、胃酸、晕动病、精神抑郁。药物也可以提高情绪、血压或促进内分泌。药物可以对抗传染性疾病，破坏肠道蠕虫，或作为其他药物的毒性解毒剂。抗肿瘤药物和放射性药物可以一起发挥作用攻击癌细胞。

药物可用于诊断糖尿病、肝功能疾病、肺结核、怀孕，或者它们可以用来补充身体缺乏的抗体、维生素、激素、电解质、蛋白质、酶或血液。药物可用于预防麻疹、脊髓灰质炎、怀孕或者可以保胎、延长生命。

当今有效药用制剂的出现代表了人类最伟大的科学成就之一。难以想象我们的文明如果缺乏这些显著有效的药剂。通过他们的使用，许多历史上困扰人类的疾病，像天花和小儿麻痹症，现在几乎绝种了。疾病如糖尿病、高血压、抑郁症和精神疾病通过现代药物得到有效控制。今天的手术几乎不可能没有麻醉剂、止痛剂、抗生素、输血、静脉输液和营养。

2. 药物的生物利用度

药物的理化性质决定着药物的吸收潜能，但是，剂型的性质（部分依赖于设计和制作工艺）也在很大程度上决定药物的生物利用度，同一药物的不同配方所存在的生物利用度差异具有临床意义。因而，药物产品的等效概念在临床决策中也很重要。化学等值是指药物制品中含有等量的同一主药并符合现行法定标准，而其中的非活性成分则可以不等。生物等效是指将化学等值的药品以同样的给药方案给予同一个体，在血液和组织中出现相等的浓度。治疗等效是指几个药物制品以同样给药方案给予同一个体，产生本质上相同的治疗效应或毒性。生物等效制品具有治疗学等效性。

有时，尽管生物利用度不同，但仍可获得治疗学等效性。例如，青霉素的治疗指数（最大耐受剂量与最小有效量之比）如此之大，以至于由青霉素制品生物利用度差异引起的中等血浓度差异可能不会影响治疗功效和安全性。相反，对于治疗指数相对狭窄的药物来说，生物利用度的差异就很重要。

生理特点以及患者的并发症也会影响生物利用度。吸收速率很重要。这是因为，即使某药物被完全吸收，如果吸收速率太慢就不能迅速达到治疗所需的血药浓度；吸收太快，则每剂用药后又会因药物浓度高而产生毒性。

当一个药物能迅速溶解并容易穿透细胞膜时，吸收趋向于完全。但口服给药时的吸收并不总是完全的。药物在到达腔静脉之前必先沿着胃肠道下行并通过肠壁和肝脏这些通常的药物代谢部位；这样，药物在进入体循环可供测量之前就可能被代谢（首过代谢）。许多药物由于首过代谢强而生物利用度低。这些组织对这些药物（如异丙肾上腺素、去甲肾上腺素、睾酮）的代谢很完全以至它们的生物利用度实际上为零。对于那些生成活性代谢物的药物来讲，经过首过代谢的治疗上的重要性取决于药物和代谢物所引起的期望的和非期望的效应。

低生物利用度最常见于水溶性差、吸收慢的药物的口服剂型。与吸收慢和不完全的药物相比，吸收迅速而完全的药物的生物利用度影响因素更多。因此，吸收缓慢或不完全常常导致治疗学效应的不同。

药物在胃肠道内停留时间不充分是低生物利用度的常见原因。所摄入药物在整个消化道的停留时间不超过 1～2d，在小肠的停留时间也只有 2～4h，如果药物溶解不迅速或不能穿透上皮细胞膜（如药物高度解离和极性强），药物在吸收部位的停留时间就可能不充分。在这种情况下，生物利用度往往变化更大，也很低。年龄、性别、活动情况、遗传表型、应激、疾病（如胃酸缺乏、营养不良综合征）或既往胃肠手术等均能影响药物的生物利用度。

妨碍吸收的许多反应能降低生物利用度。这些反应包括络合物的形成（如四环素与多价金属离子形成络合物），被胃酸或消化酶水解（如青霉素和棕榈酸氯霉素的水解），在肠壁进行结合反应（如异丙肾上腺素的硫酸结合反应），吸附于其他药物（如地高辛和消胆胺）以及被肠道菌丛代谢。

Further Reading

A Dime a Dozen Pharmaceuticals Haven't Been Synthesized

A new study has concluded that scientists have synthesized barely one tenth of 1 percent of the potential medicines that could be made. The report, in the journal ACS Chemical **Neuroscience**, estimates that the actual number of these so-called "small molecules" could be 1 **novemdecillion** (that's 1 with 60 zeroes), 1 million billion billion billion billion billion billion, which is more than some estimates of the number of stars in the universe. Jean-Louis Reymond and Mahendra Awale explain that small molecules, which are able to cross cell walls and interact with biological molecules in the body, are prime targets for scientists who develop new medicines. Most existing medications are small molecules. The authors focused on the "chemical space" **inhabited** by all of the small molecules that could possibly exist according to the laws of physics and chemistry. Researchers have identified millions of these compounds——the ACS' Chemical Abstracts Service database contains almost 67 million substances. Reymond and Awale estimate that the molecules synthesized and tested as potential drugs so far represent less than 0.1 percent of chemical space. To aid researchers looking for new ways to prevent and treat disease, they set out to find the best ways to search for new small molecules.

    The authors discuss several ways of getting a handle on chemical space, including by the size, shape and makeup of molecules. They show how computers can help researchers efficiently narrow a search for a new drug candidate. Computer modeling of chemical interactions can help researchers find a handful of promising molecules to synthesize and test in the lab. "Small molecule drugs are essential to the success of modern medicine," the authors note, and suggest that their methods may be particularly useful for finding new pharmaceuticals that target the central nervous system.

New Words

neuroscience [ˌnʊroˈsaɪəns] *n.* 神经系统科学

novemdecillion [ˌnovəmdɪˈsɪljən] *adj.* 10 的 114 次幂的；10 的 60 次幂的

inhabited [ɪnˈhæbɪtɪd] *v.* 占据

参考文献

[1] Guillaume Taglang, David B. Jackson. Use of "big data" in drug discovery and clinical trials. Gynecologic Oncology. 2016(1):17-23.

[2] Jean-Louis Montastruc, Emmanuelle Bondon-Guitton, et al, Pharmacovigilance, risks and adverse effects of self-medication. Thérapie. 2016(2):257-262.

[3] Gary Walsh. Biopharmaceuticals, biochemistry and biotechnology. Wiley. 2003.

[4] Mahendra Awale, Jean-Louis Reymond. Cluster analysis of the DrugBank chemical space using molecular quantum numbers. Bioorganic & Medicinal Chemistry. 2012, 20:5372-5378.

Chapter 2　Oral Preparation　口服制剂

1. Dosage forms application

　　Drugs are most frequently taken by oral administration. Although a few drugs taken orally are intended to be dissolved within the mouth, the vast majority of drugs taken orally are **swallowed**. Of these, the majority are taken for the systemic drug effects that result after absorption from the various surfaces along the gastrointestinal tract. A few drugs are swallowed for their local action within the confines of the gastrointestinal tract, made possible by their insolubility and/or poor absorbability from this route.

　　Compared with alternate routes, the oral route is considered the most natural, uncomplicated, convenient, and safe means of administering drugs. Disadvantages of the oral route include slow drug response (when compared with **parenterally** administered drugs) chance of irregular absorption of drugs, depending upon such factors as constitutional make-up; the amount or type of food present within the gastrointestinal tract; and the destruction of certain drugs by the acid reaction of the stomach or by gastrointestinal enzymes. Perhaps the most notable examples of the latter are the various preparations of insulin, all of which must be administered parenterally due to the destruction of this hormonal protein substance by the proteolytic enzymes of the gastrointestinal tract. The uncertainty of maintenance of the prescribed dosage regimen when the medication is in the hands of the patient is a distinct possibility, and undoubtedly many instances of overdosage or

New Words and Expressions

swallow ['swɒləʊ]
vi. 吞下；咽下

parenteral [pə'rentərəl]
adj. 不经肠道的；注射用药物的

underdosage with self-administered drugs result.

Although error in dosage is a disadvantage inherent in all types of self-administered drugs, not only the more common oral types, there is no completely effective alternative. Adequate guidance in the proper use of drugs by the patient must be provided by the physician and pharmacist.

Drugs are administered by the oral route in a variety of pharmaceutical forms, each with inherent therapeutic advantages that result in their selective use by physicians. The most popular forms are tablets, capsules, suspensions and various pharmaceutical solutions. Briefly, tablets are solid dosage forms prepared by compression or molding and contain medicinal substances with or without suitable diluents, disintegrants, coatings, colorants, and other pharmaceutical adjunct. Diluents are necessary in preparing tablets of the proper size and consistency. Disintegrants are used when rapid separation of the tablet's compressed ingredients is desired. This ensures prompt exposure of drug particles to the dissolution process thereby enhancing drug absorption. Tablet coatings are of several types and for several different purposes.

Some called **enteric** coatings are employed to permit safe passage of a tablet through the acid environment of the stomach where certain drugs may be destroyed to the more suitable juices of the intestines where tablet dissolution safely takes place. Other coatings are employed to protect the drug substance from the destructive influences of moisture, light, and air throughout their period of storage or to **conceal** a bad or bitter taste from the taste **buds** of a patient. Commercial tablets, because of their distinctive

enteric [en'terɪk]
adj. 肠的；肠溶的

conceal [kən'siːl]
vt. 隐藏；隐瞒
bud [bʌd] *n.* 味蕾

shapes, colors, and frequently employed monograms of company symbols and code numbers facilitate identification by persons trained in their use and serve as an added protection to public health.

Capsules are solid dosage forms in which the drug substance and such appropriate pharmaceutical adjuncts as fillers are enclosed in either a hard or a soft "shell", which is generally composed of a form of gelatin. Capsules vary in size, depending upon the amount of drug to be administered, and are of distinctive shapes and colors when produced commercially. Generally, drug materials are released from capsules faster than from tablets. Capsules of gelatin, a protein, are rapidly disfigured within the gastrointestinal tract, permitting the gastric juices to permeate and reach the contents.

Suspensions are preparations of finely divided drugs held in suspension throughout a suitable vehicle. Suspensions taken orally generally employ an aqueous vehicle. Nearly all suspensions must be shaken before using because they tend to settle. This ensures not only uniformity of the preparation but more importantly the administration of the proper dosage.

Suspensions are a useful means of administering large amounts of solid drugs that would be inconveniently taken in capsule form. In addition, suspensions have the advantage over solid dosage forms in that they are presented to the body in fine particle size, ready for the dissolution process immediately upon administration. However, not all oral suspensions are intended to be dissolved and absorbed by the body. For instance, Kaolin Mixture with Pectin an **antidiarrheal** preparation, contains suspended Kaolin, which acts in the intestinal tract

antidiarrheal ['æntidaiə'ri:l] *n.* 止泻剂

by adsorbing excessive intestinal fluid on the large surface area of its particles.

Drugs administered in solution are generally absorbed much more rapidly than those administered in solid form, since the process of dissolution is not required. Pharmaceutical solutions differ in the type of solvent employed and therefore in their fluidity characteristics. Several of the solutions frequently administered orally are elixirs, which are solutions in a sweetened hydro alcoholic vehicle and are generally more mobile than water; syrups, which generally utilize sucrose solutions as the sweet vehicle.

2. Absorption

Absorption of drugs after oral administration may occur at the various body sites between the mouth and **rectum**. In general, the higher up a drug is absorbed along the length of the **alimentary** tract, the more rapid will be its action, a desirable feature in most instances. However, because of the differences in the chemical and physical nature among drug substances and in the forms in which they are presented to the body, a given drug may be better absorbed from the environment of one position than from another, irrespective of its relative location within the alimentary tract.

The oral cavity is used on certain occasions as the absorption site for certain drugs. Physically, the oral absorption of drugs is managed by allowing the drug substance to be dissolved (if not presented as a solution) and withheld in the oral cavity with infrequent or no swallowing until the taste of the drug has dissipated. This process is accommodated pharmaceutically by providing the drug in a dissolved state (**impregnated** on a cotton swab, for instance)

rectum ['rektəm]
n. 直肠
alimentary [ælɪ'mentərɪ]
adj. 消化的

impregnate ['ɪmpregneɪt]
vt. 灌输；浸透

or as extremely soluble and rapidly dissolving uncoated tablets.

Drugs capable of being absorbed in the mouth may present themselves to the absorbing surface in a much more concentrated form than when swallowed, since drugs become progressively more diluted with gastrointestinal **secretions** and contents as they pass along the alimentary tract. When these secretions adversely affect the stability of a drug substance, its absorption as an active molecule is further decreased, and oral absorption or parenteral administration may be required. Similar alternatives may be required for drugs that are especially susceptible to metabolic degradation by the liver, since on gastrointestinal absorption, drug substances enter the portal circulation and are exposed to the **detoxification** processes of the liver.

Currently the oral or **sublingual** (beneath the tongue) administration of drugs is regularly employed for only a few drugs, with **nitroglycerin** and certain steroid sex hormones being the best examples. Nitroglycerin, a **coronary vasodilator** used in the **prophylaxis** and treatment of **angina** pectoris, is administered in the form of tiny tablets which are allowed to dissolve under the tongue, producing therapeutic effects in a few minutes after administration. The dose of nitroglycerin is so small (usually 400 mcg) that if it were swallowed the resulting dilute gastrointestinal concentration may not result in reliable and sufficient drug absorption. Even more important, however, is the fact that nitroglycerin is rapidly destroyed by the liver, thereby further reducing its chances of reaching its site of action. Many sex hormones have been shown

secretion [sɪ'kriːʃən]
n. 分泌；分泌物

detoxification [diˌtɑksɪfɪ'keʃən]
n. 解毒
sublingual [sʌb'lɪŋgwəl]
adj. 舌下的；舌下腺的
nitroglycerin ['naɪtrə'glɪsərɪn]
n. 硝酸甘油
coronary ['kɒrənrɪ]
adj. 冠的；冠状的
vasodilator [ˌveɪzəʊ'daɪleɪtə]
n. 血管扩张神经
prophylaxis [ˌprɒfɪ'læksɪs]
n. 预防
angina [æn'dʒaɪnə]
n. 心绞痛

to be absorbed materially better from sublingual administration than when swallowed. Although the sublingual route is probably an effective absorption route for many other drugs, it has not been extensively studied primarily because other routes have proven satisfactory and more convenient for the patient. Retaining drug substances in the mouth presents a psychological barrier to some, is unattractive to others because of the bitter taste of most drugs, and would be difficult to manage with certain patients, as children.

参考译文

1. 药物剂型应用

  药品最常服用的方式是口服。只有少部分药物需要在口腔里溶解服用，绝大多数药物需要口服后吞下。其中，绝大多数药物沿着胃肠道的表面吸收后发挥全身系统作用。还有一些药物由于可溶性差或者是吸收性差，它们在胃肠道内只能发挥局部作用。

  口服摄入与其他摄取方式相比，被认为是最自然、简单、方便、安全的药物服用的手段。口服路线的缺点包括药物作用慢（与非肠道给药方式相比），药物吸收效果可能不规律，这取决于药物本质组成；在胃肠道内的食物数量或类型；以及某些药物会被胃里的酸反应或胃肠道酶破坏。后者最著名的例子是胰岛素的多种制剂形式，无论是哪种形式都必须经非肠道给药，因为胰岛素本质是激素蛋白质容易被胃肠道的蛋白水解酶破坏。当药物在病人手中很有可能带来规定给药方案的不确定性，确实有许多实例是自行服药导致过剂量或少剂量。

  尽管药物剂量错误是所有自行给药方式固有的缺点，在口服方式中此缺点更常见，但是暂时也没有完全有效的方法代替。医师及药剂师都必须在正确用药方面给予患者一定的指导。

  口服摄入的药物有多种剂型，医师会根据每种剂型特有的治疗优势进行选择。最受欢迎的剂型是片剂、胶囊、混悬剂和各种药物溶液。简单地说，片剂属于固体剂型由压缩模具成型，片剂包含药用物质以及稀释剂、崩解剂、包衣剂、着色剂和其他药用辅料。稀释剂在赋予片剂适当大小，保证片剂均一性方面十分必要。崩解剂能够快速瓦解片剂中被压缩的成分，这样可以确保药物粒子及时进入溶解过程从而提高药物吸收度。不同类型的包衣剂具有不同的作用。

一些被称为肠溶衣的包衣材料是用于保护可能会被胃酸破坏的片剂安全通过胃的酸性环境，进入到肠道中再被溶解。其他包衣材料可以用来保护药物有效成分在有效期内不会受到水分、光照以及空气的破坏性影响，或者是为了不让患者感到药物的苦味及其他不好的味道。商品片剂，因其独特的形状、颜色和经常使用的公司符号和代码数字便于被大众识别使用，提供给公众健康保护。

胶囊剂属于固体剂型，它的药物和辅料填充于封闭的硬或软"壳"中，这种壳通常由明胶制成。胶囊大小不同，取决于药物所需量，商业化生产通常会具有独特的形状和颜色。一般来说，药物从胶囊中释放的速度比从片剂快。明胶胶囊，实质是一种蛋白质，在胃肠道中迅速变形，允许胃液渗透和内容物接触。

混悬液制剂是难溶性固体药物以微粒状态分散于分散介质，介质一般采用水溶液。几乎所有混悬液使用之前必须摇晃，因为它们容易发生沉降。这样做不仅确保药物溶液均一性，更重要的是保证患者服用剂量准确。

混悬液在需要服用大量固体药物时是一个很有效的剂型，能够克服胶囊在大量药物方面的不便。此外，混悬液剂型在一些方面优越于固体剂型，它们进入体内呈现的是小微粒状，能够立即进入溶解过程。然而，并非所有的混悬液都会在体内溶解吸收。例如，高岭土果胶合剂，用于止泻，包含悬浮高岭土，依靠这种较大表面积的微粒在肠道吸附过多的肠道液体。

药物在溶液中通常比在固体形式中吸收更迅速，因为不需要崩解溶解的过程。药物溶液在不同溶剂中有不同的流动特性。几种经常应用于口服药液的溶液是有甜味的含酒精水溶液，它们与水比流动性更好；糖浆通常利用蔗糖溶液作为载体。

2. 吸收

口服后药物的吸收可能发生在身体从嘴到直肠间的不同位点。一般来说，药物沿着消化道长度吸收得越好，药物发挥作用越迅速，在大多数情况下能得到满意的效果。然而，由于药物物质不同的化学和物理性质以及不同的形式，服用的药物可能会在某一个位置的吸收优于另一个位置，无论其在消化道内的相对位置如何。

口腔在某些情况下成为某些药物的吸收部位。从生理上讲，药物的口服吸收应当要求药物物质能够溶解（如果不是溶液形式）并且在药物味道消失之前它能够在口腔保留或还没有吞咽。这个过程的实现需要使药物处于可溶状态（例如用棉签浸透药物溶液）或采用具有极可溶性和快速溶解性的未包衣片剂。

药物的口腔吸收可能会比吞下后吸收浓度更高，因为药物吞下后沿着消化道被肠胃分泌物和内容物所稀释。如果这些分泌物影响药物稳定性，其活性分子的吸收会进一步下降，并可能需要口腔吸收或注射途径用药。类似这种特别容易被肝脏代谢降解，从胃肠道吸收后物质会进入门脉循环和肝脏的解毒过程的药物，需要采用这种方式用药。

目前口腔或舌下给药经常用于少数药物，硝酸甘油和某些类固醇性激素是最好的例子。硝酸甘油，冠状血管舒张药，用于预防和治疗心绞痛，微小片剂形式，可以溶解在舌下，口服几分钟后即能产生治疗效果。硝酸甘油的剂量太小（通常400μg），如果吞下被肠胃稀释，浓度可能不会产生可靠有效的药物吸收。更重要的是硝酸甘油会被肝脏破坏，从而进一步减少其到达病灶发挥药效的机会。许多性激素显示从舌下吸收比吞服形式更好。虽然舌下路线可能对许多其他药物来说是一个有效的吸收途径，它并没有被广泛研究主要是因为其他路线证明更令患者满意和更方便患者服用。在口腔内保持药物对一些人来说存在心理障碍，是没有吸引力的，因为大多数药物具有苦味，对于某些病人比如孩子很难接受。

Further Reading
Prescription Patches

In 1981, the motion patch was invited to prevent motion sickness. It was an amazing thing, to the world, that this little patch could prevent such **wretchedness**. Now, we have patches for everything from **arthritis** to headaches to helping us stop smoking. Are prescription patches really better than the pills we were formerly prescribed?

The concept is that a patch per day, or even per week, is much easier to remember than several pills each and every day. Prescription patches have small problems, though, like failing to stick well or being nearly impossible to remove. Also, many people break out and become **irritated** around the area where the patch is worn.

In addition to the convenience of the patch there's less irritation of the stomach and lower intestines from taking so many pills. But, the patch will not help for those who are allergic to the pill version of the same product.

Certain people should avoid particular patches. For instance, smokers, those with a history of blood **clots** or **migraines**, and those with high blood pressure should stay away from the **contraceptive** patch. Women who are breast feeding, or pregnant, and those with heart disease should avoid the nicotine replacement patch.

A pain patch, such as Duragesic, is by prescription only. The patch gives much relief to constant pain, like the pain of a cancer patient. For slightly less pain, such as **osteoarthritis**, Lidoderm is prescribed for localized pain. In lab tests the 72 hour Duragesic patch was reportedly as effective as IV **morphine**.

New Words
prescription [prɪ'skrɪpʃən] n. 药方；指示；惯例 adj. 凭处方方可购买的
patch [pætʃ] n. 贴片

wretchedness ['rɛtʃɪdnɪs] *n.* 可怜；悲惨；不幸
arthritis [ɑː'θraɪtɪs] *n.* 关节炎
irritated ['ɪrɪteɪtɪd] *adj.* 恼怒的，生气的
clot [klɒt] *n.* 凝块
migraine ['miːgreɪn] *n.* 偏头痛
contraceptive [kɒntrə'septɪv] *n.* 避孕用具；避孕剂 *adj.* 避孕的
osteoarthritis [ˌɒstɪəʊɑː'θraɪtɪs] *n.* 骨关节炎
morphine ['mɔːfiːn] *n.* 吗啡

参考文献

[1] Min H. Wu, Deborah Bartz, et al. Trends in direct-to-consumer advertising of prescription contraceptives. Contraception. 2016(5):398–405.

[2] Jean-François Etter. Short-term smoking cessation rates may be higher when people purchase nicotine gum and patches over the counter compared to obtaining treatments on prescription. Evidence-based Healthcare. 2002(4):174–175.

[3] Ana P. Ferreira, Dolapo Olusanmi, et al. Use of similarity scoring in the development of oral solid dosage forms. 2015.

[4] Gary Walsh. Biopharmaceuticals, biochemistry and biotechnology. Wiley. 2003.

Unit 6　Production of Dosage Forms　药物制剂

Chapter 3　Parenteral Administration　注射用药

1. Introduction

The first official injection (morphine) appeared in the British **Pharmacopoeia** (BP) of 1867. It was not until 1898 when cocaine was added to the BP that sterilization was attempted. In this country, the first official injections may be found in the National Formulary (NF), published in 1926.

Parenteral administration of drugs by **intravenous** (IV), **intramuscular** (IM), or **subcutaneous** (SC) routes is now an established and essential part of medical practice. Advantages for parenterally administered drugs include the following: rapid onset, predictable effect, predictable and nearly complete bioavailability, and avoidance of the gastrointestinal (GI) tract and, hence, the problems of variable absorption, drug inactivation, and GI distress. In addition, the parenteral route provides reliable drug administration in very ill or **comatose** patients.

The pharmaceutical industry directs considerable effort toward maximizing the usefulness and reliability of oral dosage forms in an effort to minimize the need for parenteral administration. Factors that contribute to this include certain disadvantages of the parenteral route, including the frequent pain and discomfort of injections, with all the psychological fears associated with "the **needle**", plus the realization that an incorrect drug or dose is often harder or impossible to counteract when it has been given parenterally (particularly intravenously), rather than orally.

In recent years, parenteral dosage forms, especially

New Words and Expressions

pharmacopoeia
[ˌfɑrməkə'piə]
n. 药典

intravenous [ˌɪntrə'viːnəs]
adj. 静脉内的

intramuscular [ˌɪntrə'mʌskjʊlə]
adj. 肌肉的

subcutaneous [ˌsʌbkjuː'teɪnɪəs]
adj. 皮下的

comatose ['kəʊmətəʊs]
adj. 昏迷的；昏睡状态的

needle ['niːdəl]
n. 针

IV forms, have enjoyed increased use. The reasons for this growth are many and varied, but they can be summed up as ① new and better parenteral administration techniques, ② an increasing number of drugs that can be administered only by a parenteral route, ③ the need for simultaneous administration of multiple drugs in hospitalized patients receiving IV therapy, ④ new forms of nutritional therapy, such as intravenous lipids, amino acids, and trace metals, and ⑤ the extension of parenteral therapy into the home.

Many important drugs are available only as parenteral dosage forms. Notable among these are numerous biotechnology drugs, insulin, several cephalosporin antibiotic products, and drugs such as **heparin**, **protamine**, and **glucagon**. In addition, other drugs, such as **lidocaine** hydrochloride and many anticancer products, are used principally as parenterals.

2. Routes of parenteral administration

The major routes of parenteral administration of drugs are subcutaneous (SC), intramuscular (IM), and intravenous (IV). Other more specialized routes are **intrathecal**, **intracisternal**, **intra-arterial**, **intraspinal**, **intrapleural**, and **intradermal**. The intradermal route is not typically used to achieve systemic drug effects.

(1) The subcutaneous route Lying immediately under the skin is a layer of fat, the superficial fascia, which lends itself to safe administration of a great variety of drugs, including vaccines, insulin, **scopolamine**, and epinephrine. The injection site may be massaged after injection to facilitate drug absorption. Drugs given by this route will have a slower onset of action than by the IM or IV routes, and total absorption may also be less.

heparin ['hepərɪn]
n. 肝素
protamine ['prəʊtəmiːn]
n. 鱼精蛋白
glucagon ['gluːkəgɒn]
n. 胰高血糖素
lidocaine ['lɪdəʊkeɪn]
n. 利多卡因
intrathecal [ˌɪntrə'θiːkəl]
adj. 囊内的
intracisternal [ˌɪntrəkɪstənl]
adj. 脑池内的
intra-arterial [ˌɪntrɑː'tɪərɪəl]
n. 动脉内的
intraspinal [ˌɪntrə'spaɪnəl]
adj. 脊柱内的
intrapleural [ˌɪntrəpl'jʊərəl]
adj. 胸膜内的
intradermal [ˌɪntrə'dɜːməl]
adj. 皮肤内的
scopolamine [skəʊ'pɒləmiːn]
n. 莨菪碱

(2) The intramuscular route   The IM route of administration is second only to the IV route in rapidity of onset of systemic action. Injections are made into the striated muscle fibers that lie beneath the subcutaneous layer. The major clinical problem arising from IM injections is muscle or **neuron** damage, the injury normally resulting from faulty technique, rather than the medication. Most injectable products can be given intramuscularly, with a normal onset of action from 15 to 30 minutes. As a result, there are numerous dosage forms available for this route of administration: solutions, oil-in-water (o/w) or water in-oil (w/o) emulsions, suspensions (aqueous or oily base), **colloidal** suspensions, and reconstitutable powders. Those product forms in which the drug is not fully dissolved generally result in slower, more gradual drug absorption, a slower onset of action, and sometimes longer lasting drug effects.

(3) The intravenous route   Intravenous medication is injected directly into a vein either to obtain an extremely rapid and predictable response or to avoid irritation of other tissues. This route of administration also provides maximum availability and assurance in delivering the drug to the site of action. However, a major danger of this route of administration is that the rapidity of absorption makes effective administration of an antidote very difficult, if not impossible, in most instances. Care must often be used to avoid administering a drug too rapidly by the IV route because irritation or an excessive drug concentration at sensitive organs such as the heart and brain (drug shock) can occur. The duration of drug activity is dependent on the initial dose and the distribution, metabolism, and

neuron ['njʊərɒn]
n. [解剖] 神经元

colloidal [kə'lɒɪdəl]
adj. 胶体的；胶质的

excretion properties (pharmacokinetics) of the drug. The IV infusion of large volumes of fluids (100~1000mL) has become increasingly popular. This technique, called **venoclysis**, utilizes products known as large-volume parenterals (LVPs). It is used to supply electrolytes and nutrients, to restore blood volume, and to prevent tissue dehydration.

3. The preparation of parenteral products

Once the formulation for a particular parenteral product is determined, including the selection of the proper solvents or vehicles and additives, the production pharmacist must follow rigid aseptic procedures in preparing the injectable products. In most manufacturing plants the area in which parenteral products are made is maintained bacteria-free through the use of ultraviolet lights, a filtered air supply, sterile manufacturing equipment, such as **flasks**, connecting tubes, and **filters**, and sterilized work clothing worn by the personnel in the area.

In the preparation of parenteral solutions, the required ingredients are dissolved according to good pharmaceutical practice either in water for injection, in one of the alternate solvents, or in a combination of solvents. The solutions are then usually filtered until sparkling clear through either **sintered** glass, porcelain, hard filter paper, or most commonly through a membrane-type filter. After filtration, the solution is transferred as rapidly as possible and with the least possible exposure into the final containers. The product is then sterilized, preferably by **autoclaving**, and samples of the finished product are tested for sterility and **pyrogens**. In instances in which sterilization by autoclaving is impractical due to the nature of the

---

venoclysis [vi'nɔklisis]
n. 静脉输注

flask [flɑːsk]
n. 烧瓶；长颈瓶
filter ['fɪltə]
n. 滤波器；过滤器

sintered ['sɪntəd]
adj. 烧结的；热压结的

autoclaving [,ɔːtəu'kleiviŋ]
n. 高压灭菌法
pyrogen ['paɪrədʒən]
n. 热原质

ingredients, the individual components of the preparation that are heat of moisture labile may be sterilized by other appropriate means and added aseptically to the sterilized solvent or to a sterile solution of all of the other components sterilizable by autoclaving.

Suspensions of drugs intended for parenteral use may be prepared by reducing the drug to a very fine powder with a ball mill, **micronizer**, colloid mill, or other appropriate equipment and then suspending the material in a liquid which it is insoluble. It is frequently necessary to sterilize separately the individual component of a suspension before combining then, as frequently the integrity of a suspension is destroyed by autoclaving. Autoclaving of a parenteral suspension may alter the viscosity of the product, thereby affecting the suspending ability of the vehicle, or change the particle size of the suspended particles, thereby altering both the pharmaceutic and the therapeutic characteristics of the preparation. If a suspension remains unaltered by autoclaving, this method is generally employed to sterilize the final product. Since parenterally administered emulsions, which are dispersions or suspensions of a liquid throughout another liquid, are generally destroyed by autoclaving, an alternate method of sterilization must be employed for this type of injectable.

Some injections are packaged as dry solids rather than in conjunction with a solvent or vehicle due to the instability of the therapeutic agent in the presence of the liquid component. These dry powdered drugs are packaged as the sterilized powder in the final containers to be reconstituted with the proper liquid prior to use, generally to

micronizer
['maɪkrənaɪzər]
n. 超微粉粉碎机

form a solution or less frequently a suspension. The method of sterilization of the powder may be dry heat or another method. It is appropriate for the particular drug involved.

In certain instances, a liquid is packaged along with the dry powder for use at the time of reconstitution. This liquid is sterile and may contain some of the desired pharmaceutical additives as the buffering agents. More frequently, the solvent or vehicle is not provided along with the dry product, but the labeling on the injection generally lists suitable solvents. Sodium chloride injection or sterile water for injection is perhaps the most frequently employed solvents used to reconstitute dry-packaged injections. The dry powders are packaged in containers large enough to permit proper shaking with the liquid component when the latter is aseptically injected through the container's rubber or plastic closure during its reconstitution. To facilitate the dissolving process, the dry powder is prevented from caking upon standing by the appropriate means including its preparation by lyophilization. Powders so treated form a honeycomb, **lattice** structure that is rapidly penetrated by the liquid, and solution is rapidly affected because of the large surface area of powder exposed.

A recent innovation in the provision of the antibiotic **cefazolin** sodium to the hospital pharmacy, has been to package the intravenous antibiotic solution in the frozen state. When **thawed**, the solution is stable for 24 hours at room temperature, or, for 10 days if stored in the refrigerator. The product is packaged in a small plastic bag for piggy-back use in intravenous administration to the patient.

lattice ['lætis]
n. [晶体] 晶格

cefazolin ['sefəzəlin]
n. 头孢唑啉
thaw [θɔː]
v. 解冻

Containers for injections, including the closures, must not interact physically or chemically with the preparation so as to alter its strength or efficacy. If the container is made of glass, it must be clear and colorless or of a light amber color to permit the inspection of its contents. The type of glass suitable and preferred for each parenteral preparation is usually stated in the individual monograph. Injections are placed either in single-dose containers or in multiple-dose containers.

Single-dose containers, commonly called **ampules** or ampoules, are sealed by fusion of the glass container under aseptic conditions. The glass container is made so as to have a neck portion that may be easily separated from the body of the container without fragmentation of the glass. After opening, the contents of the ampule may drawn into a syringe with a **hypodermic** needle. Once opened, the ampule cannot be resealed, and any unused portion may not be contained and used at a later time, since the contents would have questionable sterility. Some injectable products are packaged in prefilled syringes, with or without special administration devices.

One of the prime requisites of solution for parenteral administration is clarity. They should be sparkling clear and free of all particulate matter, that is, all of the mobile, undissolved substances which are unintentionally present. Included are such contaminants as dust, cloth fibers, glass fragments, material leached from the glass or plastic containers or seals, and any other material which may find its way into the product during its manufacture or administration, or develop during storage.

ampule
['æmpul]
n. 安瓿

hypodermic
[haɪpəʊ'dɜːmɪk]
n. 皮下注射

In order to prevent the entrance of unwanted particles into parenteral products, a number of precautions must be taken during the manufacture, storage, and use of the products. During manufacture, for instance, the parenteral solution is usually final filtered before being placed into the parenteral containers. The containers are carefully selected to be chemically resistant to the solution being added and of the highest available quality to minimize the chances of container components being leached into the solution. It has been recognized for some time, that some of the particulate matter found in parenteral products is generated from leached material from the glass or plastic containers. Once the container is selected for use, it must be carefully cleaned to be free of all extraneous matter. During container-filling, extreme care must be exercised to prevent the entrance of air-borne contaminant or other contaminants into the container. The provision of filtered and directed air flow in production areas is useful in reducing the likelihood of contamination. **Laminar flow hoods** have been developed which allow for the draft-free flow of clean filtered air over the work area. These hoods are commonly found in the hospital setting for both the manufacture and the incorporation of additives into parenteral and **ophthalmic** products. The personnel involved in the manufacture of parenterals must be made acutely aware of the importance of cleanliness and aseptic techniques. They are provided uniforms made of monofilament fabrics that do not shed lint. They wear face hoods, caps, gloves, and disposable shoe cores to prevent contamination.

After the containers are filled and **hermetically**

Laminar flow hoods ['læmɪnə fləʊ hʊds]
n. 层流净化罩

ophthalmic [ɒfˈθælmɪk]
adj. 眼睛的，眼科的

hermetically [hɜːˈmetɪkəli]
adv. 密封地，不透气地

sealed, they are visually or automatically inspected for particulate matter. Usually an inspector passes the filled container past a light source with a black back-ground to observe for mobile particles. Particles of approximately 50pm in size may be detected in this manner. Reflective particles, such as fragments of glass, may be visualized in smaller size, about 25pm in size. Other methods are used to detect particulate matter smaller than that which may be detected by the unaided eye including microscopic examinations as well as the use of **sophisticated** equipment as the Coulter Counter which electronically counts particles present in a sample presented to it. Once past the inspection following production the product may be labeled. Prior to its use, however, the pharmacist should inspect each parenteral solution dispensed for evidence of particulate matter.

sophisticated
[sə'fɪstɪkeɪtɪd]
adj. 复杂的；精致的

参考译文

1. 简介

第一个官方正式的注射剂（吗啡）出现在1867年的英国药典（BP）。直到灭菌工艺被接受认可，可卡因才被添加到1898年的BP中。在美国，第一个官方注射剂应该是出现在1926年出版的国家处方集（NF）中。

注射用药物分为静脉注射（IV）、肌肉注射（IM）或皮下注射（SC）途径，现在成为必要的医疗实践的一部分。注射用药物的优势包括以下几方面：快速起效、可预测效果、可以预见并几乎完整的生物利用度，以及避免经过胃肠道（GI），因此，吸收度多变、药物失活、胃肠道不适等问题也可以避免。此外，注射用药为病重或昏迷的病人提供了可靠的用药途径。

制药行业付出相当大的努力来最大化口服剂型的实用性和可靠性，以尽可能减少注射用制剂。造成这一现象的因素包括注射用药的某些缺点，频繁注射的疼痛和不适，对"针"所产生的心理恐惧，加上意识到当进行注射用药的给药方式（特别是静脉注射），如果药物或其剂量使用错误，通常难以或不可能有机会改正。

近年来，注射用药物的剂型，特别是静脉注射增加使用。这种增加的原因是多种多样的，但它们可以被概括为① 新的和更好的注射用技术；② 越来越多的药物只能由注射用药的途径给药；③ 住院病人需要同时服用多种药物时接受输液治疗；④ 新形式的营养治疗，如静脉注射脂质、氨基酸、微量金属；⑤ 注射用药扩展到家庭内使用。许多重要的药物只能作为注射用药物的剂型。很多生物药物采用注射剂型，其中典型是胰岛素、头孢菌素抗生素产品、肝素、鱼精蛋白、胰高血糖素。此外，其他药物如盐酸利多卡因和许多抗癌产品，主要使用注射方式。

2. 注射形式

注射用药物的主要途径是皮下、肌肉、静脉。其他更专业的途径是鞘内、脑池内、动脉内、脊柱内、硬脑膜内和皮内。皮内注射途径通常不用于实现药物发挥系统性作用。

（1）皮下注射　皮下注射是位于皮肤下的一层脂肪，又称浅筋膜，它利于安全服用各种各样的药物，包括疫苗、胰岛素、莨菪碱和肾上腺素。注射部位注射后可能需要按摩促进药物的吸收。药物通过这条途径会比 IM 或 IV 途径起效较慢的，总吸收也可能更少。

（2）肌肉注射　IM 给药途径产生系统作用速度仅次于 IV 途径。注射剂注射在皮下层的横纹肌纤维。IM 注射途径所引起的临床问题主要是肌肉或神经损伤，损伤通常是由于错误的手法，而不是药物本身。大多数可注射用产品都可以肌内注射，在 15min 到 30min 内发挥作用。因此，有许多可用此种途径注射的剂型：溶液、水包油（o/w）或油包水（w/o）乳剂（w/o）乳剂、混悬剂（水或油为基质）、胶体混悬液、可溶性粉末。这些形式产品中的药物没有完全溶解通常导致更慢、更平缓的药物吸收，起效较慢，有时药物作用也更持久。

（3）静脉注射　静脉注射药物直接注入静脉既可以获得极快的和可预测的反应又可以避免刺激其他组织。这种给药途径可以提供最大的生物利用度，并保证药物直达病灶。然而，在大多数情况下这种给药途径的主要危险是吸收的速度过快，导致如果想要解毒解药效会非常困难。必须经常使用护理手段来避免静脉用药过快，因为在敏感器官，如心脏和大脑的刺激或过量药物浓度很容易出现。药物作用的持续时间取决于药物初始剂量和其分布、代谢和排泄（药物动力学）的性质。大量的静脉输注液体（100～1000mL）已越来越受欢迎。这种技术称为静脉输注，使用的产品为大容量注射药物（LVPs）。它用于电解质和营养供应，恢复血容量，防止组织脱水。

3. 注射剂的制备

一旦注射用制剂产品的处方被确定下来，包括选择适当的溶剂或介质和添加剂，生产药剂师必须遵循严格的无菌操作程序来制备注射制剂。大多数制造工厂是通过紫外灯、空气净化系统、无菌生产设备，如玻璃瓶、连接管、过滤器、无菌工作服等的使用来保证注射制剂的生产环境处于无菌状态。

制备注射用药的过程中，根据制剂配制管理规范的要求，所需的原料应该溶解在注射用水、替代溶剂或组合溶剂中。然后溶剂通常通过烧结玻璃、瓷器、滤纸，或者通常更多的是通过膜式过滤器过滤到很干净。过滤后，溶剂尽可能迅速转移并且尽量降低暴露的可能性进入到最终的容器。然后对产品进行灭菌，最好是通过高压灭菌法，最终对成品的样本进行无菌检测和热原检查。实例中由于成分的性质导致通过高压灭菌法灭菌有时不可行，制备的个别成分可能对饱和热蒸汽不稳定，可以采用其他适当的方法对其进行消毒灭菌，再运用无菌手段将其添加到无菌溶剂中，或者将其余成分通过高压灭菌法灭菌后与该成分混合。

注射用悬浮液制剂的制备，可以通过球磨机、超微粉粉碎机、胶体磨或其他适当的设备将药物粉碎成微细粉，然后将药粉分散于不相溶的溶剂中。经常需要单独消毒各种分散成分然后再结合到一起，悬浮剂的完整性通常容易被高压灭菌法破坏。高压灭菌法可能会改变注射用悬浮物质产品的黏度，从而影响溶剂的悬浮能力，或改变悬浮粒子的粒径，从而改变制剂的性质和治疗特点。如果一个悬浮液没有被高压灭菌法改变，高压灭菌法就可以被用来消毒最终的产品。注射用乳剂，是将一种液体分散在另一种不相容的液体中，通常会被高压灭菌法破坏，必须使用另一种替代灭菌法对这一类型的制剂进行灭菌。

一些注射剂由于在液体中不稳定，因此包装成干燥固体状态，而不是与液体溶剂或介质直接结合。这些干燥的药物粉末在最终包装容器中是通过冻干法制得的，使用前用适宜的水进行再溶解形成溶液剂或悬浮液。粉末的灭菌可能会采用干燥加热或其他的方法。这适用于一些特殊药物。

在某些情况下，液体会随着干粉一起包装以备使用。这种液体是无菌的，可能包含一些所需的药用添加剂作为缓冲试剂。更常见的情况是，溶剂或介质不会和冻干产品包装在一起，但通常注射剂标签上会列出合适的溶剂。氯化钠注射液或注射用无菌水也许是最经常使用的溶剂用于重新溶解冻干剂。干燥的粉末被包装在空间足够大的容器中，以允许适当的震动来混合粉末和液体，液体是通过容器的橡胶或塑料密封塞无菌注射入容器中。为了促进溶解过程，干粉通过适当的手段包括冻干法阻止粉末粘结在一起，粉末被处理形成一个蜂窝晶格结构能够被液体迅速渗透，并且由于粉末的大表面积暴露，溶液被快速影响。

最近的创新是应用于医院药剂的抗生素头孢唑啉钠，它被制成用于静脉使用的抗生素冷冻溶液。当解冻后，溶液在室温下 24h 内都是稳定的，或者在冷冻状态能够放置 10d。产品包装在小的塑料包中用于患者背驮式手术的静脉注射。

注射剂的容器，包括瓶塞，一定不能在物理或化学性质方面影响药物的疗效。如果容器的材质是玻璃，玻璃一定要干净、无色，或呈现轻微琥珀色以便于内容物的检查。玻璃的类型应当与不同注射剂的制备相适宜，可以参见药典各论。注射剂可以单剂量包装也可以多剂量包装。

单剂量包装容器，通常称为安瓿瓶，它们在无菌条件下熔封。安瓿瓶含有瓶颈部分能够方便瓶口和瓶身分离，而不用破碎玻璃瓶。打开后，安瓿瓶中的液体被针头吸入到注射器中。一旦打开，安瓿瓶不能再封口，而且任何未被使用的部分都不能再回收使用，因为可能存在无菌问题。一些可注射产品包装在配备有特殊的服药装置的载药注射器中。

注射剂对溶液最首要的要求就是澄明度。溶液应当是澄明的不含任何悬浮微粒的，也就是说，所有的可移动微粒，包括无意间引入的不溶性物质。这些污染物质包括比如灰尘、织物纤维、玻璃碎片、从玻璃或塑料材质的容器或密封圈过滤出来的物质，以及在生产或服用以及储存过程引入的其他物质。

为了阻止不希望的小颗粒进入到注射剂当中，在生产、储存及使用药品过程中有大量的预防措施需要采取。在制造过程中，比如注射溶液在最终进入到包装容器前必须要经过过滤，包装容器一定要选择对药液化学性质稳定的，过滤容器的玻璃或塑料介质不会对药物溶液性质发生改变。一段时间以来已经认识到，大部分溶液中出现的悬浮微粒都来源于过滤介质。一旦选择了使用的容器，一定要清洗干净，防止外来杂质。在容器灌装过程中，一定要非常的仔细防止空气中的污染物或其他污染物进入到容器中。在生产环境中提供过滤过的并且控制空气流向的气流在减少污染方面很有用处。层流罩的发展出现能够在工作区域提供流动清洁的过滤空气。这种在医院常见的层流罩在注射剂以及眼用制剂的生产灌装中被使用。注射剂生产的工作人员必须清醒地意识到清洁以及无菌技术的重要性。他们被提供了统一的工作服由没有线头的单丝面料制成，他们穿戴面罩、头套、手套、一次性鞋套来防止污染。

在容器灌装和密封后，要接受视觉上或自动的颗粒物检查。通常一个检查员将灌装容器置于一个黑色背景的光源下观察可移动粒子。直径大约为50pm 的颗粒可以采用这种方式检测，可反光粒子，比如玻璃碎片，可能会更小一些，直径大约在25pm。其他用于检测大小小于肉眼所能检测到的颗粒物的方法，包括显微观察以及使用复杂电子设备的库尔特计数器，来计数存在于样本中的粒子。产品一旦通过检查，需要贴标签。在使用它之前，无论怎么样，药剂师应该检查每个注射剂溶液中是否有悬浮粒子的存在。

Further Reading
Sugar can Reduce Infant Injection Pain

Babies should be given something sugary before a **jab** to reduce pain, Canadian researchers say. Experts at the University of Toronto say newborns are less likely to cry if given a few drops of a sugar solution before immunisation.

Data based on 1,000 injections suggests infants given a glucose solution are 20% less likely to cry following a jab. The research, published in Archives of Disease in Childhood, is based on clinical data from 14 studies. A team led by Dr Arne Ohlsson, of the University of Toronto, looked at data from **clinical trials** in babies up to a year old. As well as the findings related to glucose, the researchers found that between a few drops and half a teaspoon of sucrose and glucose also led to a small reduction in the amount of time a baby spent crying. The Canadian researchers, who worked in collaboration with colleagues in Australia and Brazil, concluded: "Healthcare professionals should consider using sucrose or glucose before and during immunisation."

Existing research points to the **pain-relieving** properties of sweet solutions working for babies undergoing painful procedures such as a **heel prick**. Adam Finn, professor of **pediatrics** at the University of Bristol, said: "Anything we can do to minimise the discomfort of immunisation for children is to be welcomed, and I would like to see more research in this area. "On the one hand parents are more likely to return if the experience is not distressing. But more fundamentally, children don't agree to have **vaccines**, so we need to be sure we are making it as painless as possible for them."

New Words

jab [dʒæb] *n.* 戳；猛击；注射
clinical trial ['klɪnɪk(ə)l 'traɪəl] *n.* 临床试验
pain-relieving ['peɪnrɪl'iːvɪŋ] *n.* 止痛剂
heel [hiːl] *n.* 脚后跟
prick [prɪk] *n.* 刺，扎；刺痛
pediatric [ˌpidɪ'ætrɪk] *adj.* 小儿科的

参考文献

[1] Roman Mathaes, Hanns-Christian Mahler, et al, The pharmaceutical vial capping process: Container closure systems, capping equipment, regulatory framework, and seal quality tests. European Journal of Pharmaceutics and Biopharmaceutics. 2016(99):54-64.

[2] M.M. Stark. Substance Misuse: Methods of Administration. Encyclopedia of Forensic and Legal Medicine (Second Edition). 2016:406-410.

[3] Geert J.A. Wanten. MD, et al. Parenteral approaches in malabsorption: Home parenteral nutrition. Best Practice & Research Clinical Gastroenterology. 2016(2): 309-318.

[4] Gilbert S. Banker, Christopher T. Rhodes. Modern Pharmaceutics. Marcel Dekker, Inc. 2002.

Chapter 4 Drug for External Use 外用药

1. Epicutaneous route

Drugs are administered topically, or applied to the skin, chiefly for their action at the site of application or for systemic drug effects.

In general, drug absorption via the skin is enhanced if the drug substance is in solution, if it has a favorable lipid/water partition coefficient, and if it is a nonelectrolyte. Drugs that are absorbed enter the skin by way of the pores, sweat glands, hair follicles, sebaceous glands, and other anatomic structures of the skin's surface. Since blood capillaries are present just below the **epidermal** cells, a drug that penetrates the skin and is able to traverse the capillary wall finds ready access to the general circulation.

Among the few drugs currently employed topically to the skin surface for **percutaneous** absorption and systemic action are nitroglycerin and scopolamine. Each of these drugs is available for use in the form of **transdermal** delivery systems fabricated as an adhesive disc or patch which slowly releases the medication for percutaneous absorption. Additionally, nitroglycerin is available in an ointment form for application to the skin's surface for systemic absorption. Nitroglycerin is employed therapeutically for **ischemic** heart disease, with the transdermal dosage forms becoming increasingly popular because of the benefit in patient compliance through their long-acting (24 hours) characteristics. The nitroglycerin patch is generally applied to the arm or chest, preferably in a hair-free or shaven area. The transdermal scopolamine system is also in the form of a patch to be applied to the skin;

New Words and Expressions

epicutaneous
adj. 经皮肤的

epidermal [ˌepiˈdəːməl]
adj. 表皮的；外皮的

percutaneous [ˌpɜːkjʊˈteɪnɪəs]
adj. 经皮的；经由皮肤的
transdermal [trænsˈdəːməl]
adj. 经皮的，经皮肤

ischemic [ɪsˈkimɪk]
adj. 缺血性的；局部缺血的

In this case, behind the ear. The drug system is indicated for the prevention of **nausea** and **vomiting** associated with motion sickness. The commercially available product is applied to the **postauricular** area several hours before need (as prior to an air or sea trip) where it releases its medication over a period of 3 days.

For the most part, pharmaceutical preparations applied to the skin are intended to serve some local action and as such are formulated to provide prolonged local contact with minimal absorption. Drugs applied to the skin for their local action include antiseptics, antifungal agents, anti-inflammatory agents, local **anesthetic** agents, skin **emollients**, and protectants, against environmental conditions, as the effects of the sun, wind, pests, and chemical irritants. For these purposes drugs are most commonly administered in the form of ointments and related semisolid preparations such as creams and pastes, as solid dry powders, aerosol sprays or as liquid preparations such as solutions and lotions.

Pharmaceutically, ointments, creams, and pastes are semisolid preparations in which the drug is contained in a suitable base (ointment base) which is itself semisolid and either hydrophilic or hydrophobic in character. These bases play an important role in the proper formulation of semisolid preparations, and there is no single base universally suitable as a carrier of all drug substances or for all therapeutic indications. The proper base for a drug must be determined individually to provide the desired drug release rate, staying qualities after application, and texture. Briefly, ointments are simple mixtures of drug substances in an ointment base, whereas creams

nausea ['nɔːziə]
n. 恶心，晕船
vomiting ['vɒmitiŋ]
v. 呕吐
postauricular [pəʊstɔːˈrɪkjʊlə]
adj. 耳廓后的

anesthetic [ˌænisˈθetik]
n. 麻醉剂，麻药
emollient [ɪˈmɒlɪənt]
n. 润肤剂；软化剂

are semisolid emulsions and are generally less viscid and lighter than ointments. Creams are considered to have greater **esthetic** appeal due to their nongreasy character and their ability to "**vanish**" into the skin upon rubbing. Pastes contain more solid materials than do ointments and are therefore stiffer and less penetrating. Pastes are usually employed for their protective action and for their ability to absorb serous discharges from skin **lesions**. Thus when protective rather than therapeutic action is desired, the formulation pharmacist will favor paste, but when therapeutic action is required, he will prefer ointments and creams. Commercially, many therapeutic agents are prepared in both ointment and cream form and are dispensed and used according to the particular preference of the patient and the prescribing **practitioner**.

Medicinal powders are simply intimate mixtures of medicinal substances usually in an **inert** base like talcum powder. Depending upon the particle size of the resulting blend, the powder will have varying dusting and covering capabilities. In any case, the particle size should be small enough to ensure against grittiness and consequent skin irritation. Powders are most frequently applied topically to relieve such conditions as diaper rash, chafing, and athlete's foot.

When topical application is desired in liquid form other than solution, lotions are most frequently employed. Lotions are generally suspensions of solid materials in an aqueous vehicle, although certain emulsions and even some true solutions have been designated as lotions because of either their appearance or application. Lotions may be

esthetic [iːsˈθetɪk]
adj. 审美的
vanish [ˈvænɪʃ]
vi. 消失；突然不见

lesion [ˈliːʒən]
n. 损害；身体上的伤害

practitioner [prækˈtɪʃənə]
n. 从业者，执业医生
inert [ɪˈnɜːt]
adj. 惰性的

preferred over semisolid preparations because of their nongreasy character and their increased spreadability over large areas of skin.

2. **Ocular**, **otic**, and **nasal** routes

Drugs are frequently applied topically to the eye, ear, and the mucous membranes of the nose. In these instances, ointments, suspensions, and solutions are generally employed. **Ophthalmic** solutions and suspensions are sterile aqueous preparations with other qualities essential to the safety and comfort of the patient. Ophthalmic ointments must be sterile, and also free of **grittiness**. Nasal preparations are usually solutions or suspensions administered by drops or as a fine mist from a nasal spray container. Otic, or ear preparations are usually viscid so that they have prolonged contact with the affected area. They may be employed simply to soften ear wax, to relieve an earache, or to combat an ear infection. Eye, ear, and nose preparations are not generally employed for systemic effects, and although ophthalmic and otic preparations are not usually absorbed to any great extent, nasal preparations may be absorbed, and systemic effects after the intranasal application of solution are not unusual.

ocular ['ɒkjʊlə]
adj. 眼睛的
otic ['əʊtɪk]
adj. 耳的；耳部的
nasal ['neɪzəl]
adj. 鼻的；鼻音的
ophthalmic [ɒf'θælmɪk]
adj. 眼睛的，眼科的
grittiness ['grɪtɪnɪs]
n. 砂砾

参考译文

1. 皮肤给药

药物局部使用，或应用于皮肤，主要是为了使它们能在局部使用的位置产生局部效果或发挥全身系统性效果。一般来说，如果药物溶液是非电解质，而且其中的物质有良好的脂/水分配系数，药物通过皮肤吸收会增强。药物吸收进入皮肤通过毛孔、汗腺、毛囊、皮脂腺等解剖结构的皮肤表面。由于毛细血管存在表皮细胞的下面，药物穿透皮肤可以透过血管壁进入到体循环中。

通过皮肤途径给药应用于局部作用或全身作用的药物中典型的是硝酸甘油和莨菪碱。这些药物可用于制成贴片通过透皮给药法慢慢释放药物并经皮吸收。此外，硝酸甘油软膏剂形式应用于皮肤表面也可用于全身吸收。硝酸甘油用于临床治疗局部缺血性心脏病，制成透皮给药的制剂剂型逐渐流行起来，因为它的作用长效（24h），病人的服药依从性更好。硝酸甘油贴片通常应用于手臂或胸部，最好是在背部或刮过的部位。莨菪碱也是以贴片的形式被应用到皮肤，可以贴在耳后。药物发挥作用旨在防止恶心呕吐以及晕动病。市场上可买到的产品应用于耳廓后部，几小时后发挥药效（在飞行前或坐船前使用）释放药效长达三天。

在大多数情况下，制剂应用于皮肤的目的是为局部作用发挥药效，因此配方应用最小吸收量提供持久的局部作用。药物应用于皮肤发挥局部作用包含防腐剂、抗真菌剂、抗炎药、局部麻醉剂、润肤剂、保护剂，它们可以对抗环境中的阳光、风、害虫和化学刺激物的影响。因此这种类型的制剂通常制成药膏形式，或半固体形式比如霜、糊剂、固体干粉、气溶胶喷雾，或者液体形式比如溶液剂和洗剂。

药用软膏、乳霜、糊剂属于半固体制剂，它们的药物是包含在一个合适的基质中（软膏基质），基质本身是半固体和亲水性或疏水性的。这些基质在制备半固体制剂中扮演重要角色，并没有一个单一的基质能够适合所有药物物质或适应证。适宜的基质必须提供给药物恰当的释放速率、应用后保持质量、质地。简略地说，药膏是简单地将药物物质和药膏基质混合在一起，而霜是半固体乳剂，比药膏黏性小而且更轻质。霜被认为审美吸引力更大，因为它们非油性而且通过摩擦皮肤可以吸收进去。贴比软膏含有更多的固体材料，因此更硬，渗透性稍弱些。贴通常用于起到保护作用，他们从损伤的皮肤吸收渗出液。因此，当需要保护而不是治疗作用时，药剂师更倾向于贴剂，但当治疗作用是必需时，他们会选择药膏和霜。商业上，许多治疗药物制备成药膏和霜的形式，根据病人和处方医生的特殊偏好来分发和使用。

药用粉末只是简单地把药用物质和惰性基质比如滑石粉进行均匀混合。根据混合的粒度，粉会有不同的清除和覆盖能力。在任何情况下，粒子的大小应该足够小，以防止皮肤使用时带来的刺激性和粗硬感。粉最经常在局部使用，用于舒缓尿布疹、皮肤发炎和脚气。

当局部应用所需的液体是溶液剂以外的形式，洗剂是最经常使用的。洗剂通常是固体材料悬浮在一种水溶剂中，虽然某些乳剂，甚至一些真正的溶液剂也被称为洗剂。洗剂可能优于半固体制剂，因为它们的非油性性质，并且能够增加在大面积皮肤上的覆盖性。

2. 眼睛、耳、鼻用药

药物通常还可以局部应用于眼睛、耳朵和鼻子的黏膜，在这些情况下，药膏、悬浮液和溶液剂是通常使用的剂型。眼用溶液剂和悬浮液是无菌水溶液制剂，必须具备的特质就是对患者来讲要安全舒适。眼药膏也必须无菌，也不应含有沙砾等异物。鼻用制剂通常是溶液剂或悬浮液以滴剂形式或通过喷鼻剂容器以小雾滴形式使用。耳部或者耳朵用制剂通常是半流体的保证它们可以与损伤部位长效接触。它们可以软化耳垢、减轻耳痛或抵抗耳朵感染。眼睛、耳朵、鼻子的制剂通常不用于发挥全身药效,眼和耳的制剂通常在大范围内不被吸收,鼻腔制剂可能被吸收,溶液剂应用于鼻内很少起全身作用。

Further Reading
Ocular Medications

The three primary methods of delivery of ocular medications to the eye are topical, local ocular (ie, **subconjunctival**, **intravitreal**, **retrobulbar**, intracameral), and systemic. The most appropriate method of administration depends on the area of the eye to be medicated. The conjunctiva, **cornea**, **anterior chamber**, and iris usually respond well to topical therapy. The **eyelids** can be treated with topical therapy but more frequently require systemic therapy. The **posterior** segment always requires systemic therapy, because most topical medications do not penetrate to the posterior segment. Retrobulbar and **orbital** tissues are treated systemically.

Subconjunctival or sub-Tenon's therapy, although not a true form of systemic medication administration, has the potential to increase both drug absorption and contact time. Medications both leak onto the cornea from the entry hole of injection and diffuse through the sclera into the globe. Drugs with low solubility such as **corticosteroids** may provide a repository of drug lasting days to weeks. Appropriate amounts of medication must be used. Large amounts, especially of long-acting salts, can cause a significant inflammatory reaction. For sub-Tenon's injections, 0.5 mL per site is usually safe and effective in small animals and $\leqslant$ 1 mL can be used in large animals such as horses and cows.

Retrobulbar medications are used infrequently for therapeutics. In cattle, the retrobulbar tissues can be anesthetized with local anesthetic (lidocaine/**bupivicaine**) for **enucleation** using either a Peterson block (15~20 mL) or a 4 point block of the orbit (5~10 mL/site). Whenever any medication is placed into the orbit, extreme care must be taken to ensure that the medication is not inadvertently injected into a blood vessel, the optic nerve, or one of the orbital foramen. Retrobulbar injection has a high risk of adverse effects and should not be used unless the clinician is experienced and the animal is appropriately restrained.

Systemic medication is required for posterior segment therapy and to complement topical therapy for the anterior segment. The blood-ocular barriers can limit absorption of less **lipophilic** drugs, but inflammation will initially allow greater drug concentrations to reach the site. As the eye starts to heal, these barriers will again become more effective and can limit further drug penetration. This should be considered when treating posterior segment disease, eg, **blastomycosis** in small animals with hydrophilic drugs such as **itraconazole**.

After topical administration, up to 80% of the applied drug (s) is absorbed systemically across the highly **vascularized nasopharyngeal mucosa**. Because absorption via this route by passes the liver, there is not the large first-pass metabolism seen after oral administration. Depending on the drugs used, this can result in systemic adverse effects. Topically applied β-blockers used in the treatment of glaucoma can cause heart block, **atrial tachycardia**, **congestive heart failure**, **bronchospasm**, **dyspnea**, and decreased exercise tolerance. These drugs should be used very carefully in older animals or in animals with **cardiac** or **respiratory** disease. Cushing **syndrome** can be easily induced in small or medium-sized dogs with chronic use of potent topical steroids.

New Words

subconjunctival [sʌbkənd'ʒʌŋktaɪvl] adj. 结膜下的
intravitreal [ɪntrævɪtr'ɪɔ:l] adj. 玻璃体内的
retrobulbar [ˌretrəʊ'bʌlbə] adj. 眼球后的
cornea ['kɔːnɪə] n. 角膜
anterior [æn'tɪərɪə] adj. 前面的；先前的
chamber ['tʃeɪmbə] n. 室，膛
eyelid ['aɪlɪd] n. 眼睑；眼皮
posterior [pɒ'stɪərɪə] n. 后部；臀部
orbital ['ɔːbɪtəl] adj. 轨道的；眼窝的
corticosteroid [ˌkɔːtikəʊs'tɪrɔɪd] n. 皮质类固醇，类固醇
bupivicaine [bjʊpɪvə'keɪn] n. 丁哌卡因
enucleation [iˌnjuːkli'eɪʃən] n. 摘出术
lipophilic [ˌlɪpəʊ'fɪlɪk] adj. 亲脂性的，亲脂的
blastomycosis [ˌblæstəʊmaɪ'kəʊsɪs] n. 酿母菌病
itraconazole [ɪtrə'kənəzəʊl] n. 依曲康唑
vascularize ['væskjʊləraiz] vt. 血管化
nasopharyngeal [ˌneɪzəʊfə'rɪndʒɪəl] adj. 鼻咽的
mucosa [mjuː'kəʊsə] n. 黏膜
atrial tachycardia ['eɪtrɪəl ˌtækɪ'kɑːdɪə] n. 心房性心搏过速
congestive heart failure [kən'dʒestɪv hɑːt 'feiljə] n. 充血性心力衰竭
bronchospasm ['brɒŋkəspæzəm] n. 支气管痉挛
dyspnea [dɪsp'niːə] n. 呼吸困难
cardiac ['kɑːdɪæk] adj. 心脏的；心脏病的

respiratory [rəˈspɪrətri] *adj.* 呼吸的

syndrome [ˈsɪndrəʊm] *n.* 综合征；综合症状；并发症状

参考文献

[1] Arto Urtti, Lotta Salminen. Minimizing systemic absorption of topically administered ophthalmic drugs. Survey of Ophthalmology. 1993(6):435-456.

[2] 吴达俊. 制药工程专业英语. 北京：化学工业出版社, 2000.

[3] 胡廷熹. 药学英语. 北京：人民卫生出版社, 2007.

Unit 7 Good Manufacturing Practice for Drugs   GMP

Chapter 1 Process Validation: General Principles and Practices   工艺验证：一般原则与规范

In the following sections, we describe general considerations for process validation, the recommended stages of **process validation**, and specific activities for each stage in the product lifecycle.

1. General considerations for process validation

In all stages of the product lifecycle, good project management and good archiving that capture scientific knowledge will make the process validation program more effective and efficient. The following practices should ensure uniform collection and assessment of information about the process and enhance the accessibility of such information later in the product lifecycle.

(1) We recommend an integrated team approach to process validation that includes expertise from a variety of disciplines (e.g., process engineering, industrial pharmacy, analytical chemistry, microbiology, statistics, manufacturing, and quality assurance). Project plans, along with the full support of senior management, are essential elements for success.

(2) Throughout the product lifecycle, various studies can be initiated to discover, observe, correlate, or confirm information about the product and process. All studies should be planned and conducted according to sound scientific principles, appropriately documented, and approved in accordance with the

New Words and Expressions

process ['prəʊses]
n. 过程，进行；方法，步骤；工艺

validation [ˌvælɪ'deɪʃən]
n. 确认；验证

established procedure appropriate for the stage of the lifecycle.

(3) The terms attribute(s) (e.g., quality, product, component) and parameter(s) (e.g., process, operating, and equipment) are not categorized with respect to criticality in this guidance. With a lifecycle approach to process validation that employs risk based decision making throughout that lifecycle, the perception of criticality as a continuum rather than a binary state is more useful. All attributes and parameters should be evaluated in terms of their roles in the process and impact on the product or in-process material, and reevaluated as new information becomes available. The degree of control over those attributes or parameters should be commensurate with their risk to the process and process output. In other words, a higher degree of control is appropriate for attributes or parameters that pose a higher risk. The Agency recognizes that **terminology** usage can vary and expects that each manufacturer will communicate the meaning and intent of its terminology and categorization to the Agency.

terminology
[ˌtɜːmɪ'nɒlədʒɪ]
n. 术语，术语学

(4) Many products are single-source or involve complicated manufacturing processes. **Homogeneity** within a batch and **consistency** between batches are goals of process validation activities. Validation offers assurance that a process is reasonably protected against sources of variability that could affect production output, cause supply problems, and negatively affect public health.

homogeneity
[ˌhɒməʊdʒɪ'neɪɪtɪ]
n. 同质；同种
consistency
[kən'sɪstənsɪ]
n. 一致性；稠度；相容性

2. Stage 1——process design

Process design is the activity of defining the commercial manufacturing process that will be reflected in planned master production and control records. The goal of this stage is to design a process suitable for routine commercial manufacturing that can consistently deliver a product that meets its quality attributes.

(1) Building and capturing process knowledge and understanding Generally, early process design experiments do not need to be performed under the CGMP conditions required for drugs intended for commercial distribution that are manufactured during Stage 2 (process qualification) and Stage 3 (continued process **verification**). They should, however, be conducted in accordance with sound scientific methods and principles, including good **documentation** practices. This recommendation is consistent with ICH Q10 Pharmaceutical Quality System. Decisions and justification of the controls should be sufficiently documented and internally reviewed to verify and preserve their value for use or adaptation later in the lifecycle of the process and product.

Although often performed at small-scale laboratories, most viral **inactivation** and impurity clearance studies cannot be considered early process design experiments. Viral and impurity clearance studies intended to evaluate and estimate product quality at commercial scale should have a level of quality unit oversight that will ensure that the studies follow sound scientific methods and principles and the conclusions are supported by the data.

Product development activities provide key inputs to the process design stage, such as the intended dosage form, the quality attributes, and a general manufacturing pathway. Process information available from product development

verification
[ˌverɪfɪ'keɪʃən]
n. 确认，查证；核实
documentation
[ˌdɒkjʊmen'teɪʃən]
n. 文件，证明文件，史实，文件编制

inactivation
[inˌækti'veiʃən]
n. 失活；钝化（作用）

activities can be leveraged in the process design stage. The functionality and limitations of commercial manufacturing equipment should be considered in the process design, as well as predicted contributions to variability posed by different component lots, production operators, environmental conditions, and measurement systems in the production setting. However, the full spectrum of input variability typical of commercial production is not generally known at this stage. Laboratory or pilot-scale models designed to be representative of the commercial process can be used to estimate variability.

Designing an efficient process with an effective process control approach is dependent on the process knowledge and understanding obtained. Design of Experiment (DOE) studies can help develop process knowledge by revealing relationships, including **multivariate** interactions, between the variable inputs (e.g., component characteristics or process parameters) and the resulting outputs (e.g., in-process material, intermediates, or the final product). Risk analysis tools can be used to screen potential variables for DOE studies to minimize the total number of experiments conducted while maximizing knowledge gained. The results of DOE studies can provide justification for establishing ranges of incoming component quality, equipment parameters, and in-process material quality attributes. FDA does not generally expect manufacturers to develop and test the process until it fails.

Other activities, such as experiments or demonstrations at laboratory or pilot scale, also assist in evaluation of certain conditions and prediction

multivariate
[ˌmʌltɪ'veərɪət]
n. 多元；多变量
adj. 多元的；多变量的

of performance of the commercial process. These activities also provide information that can be used to model or simulate the commercial process. Computer-based or virtual simulations of certain unit operations or dynamics can provide process understanding and help avoid problems at commercial scale. It is important to understand the degree to which models represent the commercial process, including any differences that might exist, as this may have an impact on the relevance of information derived from the models.

It is essential that activities and studies resulting in process understanding be documented. Documentation should reflect the basis for decisions made about the process. For example, manufacturers should document the variables studied for a unit operation and the rationale for those variables identified as significant. This information is useful during the process qualification and continued process verification stages, including when the design is revised or the strategy for control is refined or changed.

(2) Establishing a strategy for process control

Process knowledge and understanding is the basis for establishing an approach to process control for each unit operation and the process overall. Strategies for process control can be designed to reduce input variation, adjust for input variation during manufacturing (and so reduce its impact on the output), or combine both approaches.

Process controls address variability to assure quality of the product. Controls can consist of material analysis and equipment monitoring at significant processing points. Decisions regarding the type and extent of process controls can be aided by earlier risk assessments, then enhanced and improved as process experience is gained.

FDA expects controls to include both examination of material quality and equipment monitoring. Special attention to control the process through operational limits and in-process monitoring is essential in two possible scenarios:

① When the product attribute is not readily measurable due to limitations of sampling or detectability (e.g., viral clearance or microbial **contamination**) or

② When intermediates and products cannot be highly characterized and well-defined quality attributes cannot be identified.

These controls are established in the master production and control records.

More advanced strategies, which may involve the use of process analytical technology (PAT), can include timely analysis and control loops to adjust the processing conditions so that the output remains constant. Manufacturing systems of this type can provide a higher degree of process control than non-PAT systems. In the case of a strategy using PAT, the approach to process qualification will differ from that used in other process designs.

The planned commercial production and control records, which contain the operational limits and overall strategy for process control, should be carried forward to the next stage for confirmation.

3. Stage 2——process qualification

During the process qualification (PQ) stage of process validation, the process design is evaluated to determine if it is capable of reproducible commercial

contamination
[kən͵tæmɪˈneɪʃən]
n. 污染，玷污；污染物

manufacture. This stage has two elements: ① design of the **facility** and qualification of the equipment and utilities and ② process performance qualification (PPQ). During Stage 2, CGMP-compliant procedures must be followed. Successful completion of Stage 2 is necessary before commercial distribution. Products manufactured during this stage, if acceptable, can be released for distribution.

(1) Design of a facility and qualification of utilities and equipment   Proper design of a manufacturing facility is required under part 211, subpart C, of the CGMP regulations on Buildings and Facilities. It is essential that activities performed to assure proper facility design and commissioning precede PPQ. Here, the term *qualification* refers to activities undertaken to demonstrate that utilities and equipment are suitable for their intended use and perform properly. These activities necessarily precede manufacturing products at the commercial scale.

Qualification of utilities and equipment generally includes the following activities:

① Selecting utilities and equipment construction materials, operating principles, and performance characteristics based on whether they are appropriate for their specific uses.

② Verifying that utility systems and equipment are built and installed in compliance with the design specifications (e.g., built as designed with proper materials, capacity, and functions, and properly connected and calibrated).

③ Verifying that utility systems and equipment operate in accordance with the process requirements in all anticipated operating ranges. This should include challenging the equipment or

facility [fə'sɪləti]
n. 设施；设备；容易；灵巧

system functions while under load comparable to that expected during routine production. It should also include the performance of interventions, stoppage, and start-up as is expected during routine production. Operating ranges should be shown capable of being held as long as would be necessary during routine production.

Qualification of utilities and equipment can be covered under individual plans or as part of an overall project plan. The plan should consider the requirements of use and can incorporate risk management to prioritize certain activities and to identify a level of effort in both the performance and documentation of qualification activities. The plan should identify the following items:

The studies or tests to use;

The criteria appropriate to assess outcomes;

The timing of qualification activities;

The responsibilities of relevant departments and the quality unit;

The procedures for documenting and approving the qualification.

The project plan should also include the firm's requirements for the evaluation of changes. Qualification activities should be documented and summarized in a report with conclusions that address criteria in the plan. The quality control unit must review and approve the qualification plan and report.

(2) Process performance qualification The process performance qualification (PPQ) is the second element of Stage 2, process qualification. The PPQ combines the actual facility, utilities, equipment (each now qualified), and the trained personnel with the commercial manufacturing

process, control procedures, and components to produce commercial batches. A successful PPQ will confirm the process design and demonstrate that the commercial manufacturing process performs as expected.

Success at this stage signals an important milestone in the product lifecycle. A manufacturer must successfully complete PPQ before commencing commercial distribution of the drug product. The decision to begin commercial distribution should be supported by data from commercial-scale batches. Data from laboratory and pilot studies can provide additional assurance that the commercial manufacturing process performs as expected.

The approach to PPQ should be based on sound science and the manufacturer's overall level of product and process understanding and demonstrable control. The cumulative data from all relevant studies (e.g., designed experiments; laboratory, pilot, and commercial batches) should be used to establish the manufacturing conditions in the PPQ. To understand the commercial process sufficiently, the manufacturer will need to consider the effects of scale. However, it is not typically necessary to explore the entire operating range at commercial scale if assurance can be provided by process design data. Previous credible experience with sufficiently similar products and processes can also be helpful. In addition, we strongly recommend firms employ objective measures (e.g., statistical metrics) wherever feasible and meaningful to achieve adequate assurance.

In most cases, PPQ will have a higher level of sampling, additional testing, and greater scrutiny of process performance than would be typical of routine commercial production. The level of monitoring and testing should be sufficient to confirm uniform product quality throughout the batch. The increased level of scrutiny, testing, and sampling should continue through the process verification stage as appropriate, to establish levels and frequency of routine sampling and monitoring for the particular product and process. Considerations for the duration of the heightened sampling and monitoring period could include, but are not limited to, volume of production, process complexity, level of process understanding, and experience with similar products and processes.

The extent to which some materials, such as column **resins** or molecular filtration media, can be re-used without adversely affecting product quality can be assessed in relevant laboratory studies. The usable lifetimes of such materials should be confirmed by an ongoing PPQ protocol during commercial manufacture.

resin ['rezɪn]
n. 树脂；松香

A manufacturing process that uses PAT may warrant a different PPQ approach. PAT processes are designed to measure in real time the attributes of an in-process material and then adjust the process in a timely control loop so the process maintains the desired quality of the output material. The process design stage and the process qualification stage should focus on the measurement system and control loop for the measured attribute. Regardless, the goal of validating any manufacturing process is the same: to establish scientific evidence that the process is reproducible and will consistently deliver quality products.

(3) PPQ protocol   A written protocol that specifies the manufacturing conditions, controls, testing, and expected outcomes is essential for this stage of process validation. We recommend that the protocol discuss the following elements:

① The manufacturing conditions, including operating parameters, processing limits, and component (raw material) inputs.

② The data to be collected and when and how it will be evaluated.

③ Tests to be performed (in-process, release, characterization) and acceptance criteria for each significant processing step.

④ The sampling plan, including sampling points, number of samples, and the frequency of sampling for each unit operation and attribute. The number of samples should be adequate to provide sufficient statistical confidence of quality both within a batch and between batches. The confidence level selected can be based on risk analysis as it relates to the particular attribute under examination. Sampling during this stage should be more extensive than is typical during routine production.

⑤ Criteria and process performance indicators that allow for a science- and risk-based decision about the ability of the process to consistently produce quality products. The criteria should include:

A description of the statistical methods to be used in analyzing all collected data (e.g., statistical metrics defining both intra-batch and inter-batch variability).

Provision for addressing **deviations** from expected conditions and handling of nonconforming

deviation [diːvɪˈeɪʃən]
n. 偏差；误差；背离

data. Data should not be excluded from further consideration in terms of PPQ without a documented, science-based justification.

⑥ Design of facilities and the qualification of utilities and equipment, personnel training and qualification, and verification of material sources (components and container/closures), if not previously accomplished.

⑦ Status of the validation of analytical methods used in measuring the process, in-process materials, and the product.

⑧ Review and approval of the protocol by appropriate departments and the quality unit.

(4) PPQ protocol execution and report
Execution of the PPQ protocol should not begin until the protocol has been reviewed and approved by all appropriate departments, including the quality unit. Any departures from the protocol must be made according to established procedure or provisions in the protocol. Such departures must be justified and approved by all appropriate departments and the quality unit before implementation.

The commercial manufacturing process and routine procedures must be followed during PPQ protocol execution. The PPQ lots should be manufactured under normal conditions by the **personnel** routinely expected to perform each step of each unit operation in the process. Normal operating conditions should include the utility systems (e.g., air handling and water purification), material, personnel, environment, and manufacturing procedures.

A report documenting and assessing adherence to the written PPQ protocol should be prepared in a

personnel [pɜːsə'nel]
n. 人事部门；(全体) 人员

timely manner after the completion of the protocol. This report should:

① Discuss and cross-reference all aspects of the protocol.

② Summarize data collected and analyze the data, as specified by the protocol.

③ Evaluate any unexpected observations and additional data not specified in the protocol.

④ Summarize and discuss all manufacturing nonconformances such as deviations, aberrant test results, or other information that has bearing on the validity of the process.

⑤ Describe in sufficient detail any corrective actions or changes that should be made to existing procedures and controls.

⑥ State a clear conclusion as to whether the data indicates the process met the conditions established in the protocol and whether the process is considered to be in a state of control. If not, the report should state what should be accomplished before such a conclusion can be reached. This conclusion should be based on a documented justification for the approval of the process, and release of lots produced by it to the market in consideration of the entire compilation of knowledge and information gained from the design stage through the process qualification stage.

⑦ Include all appropriate department and quality unit review and approvals.

4. Stage 3——continued process verification

The goal of the third validation stage is continual assurance that the process remains in a state of control (the validated state) during commercial manufacture. A system or systems for detecting unplanned departures from the process as designed is essential to accomplish this goal. Adherence to the CGMP requirements, specifically, the collection and evaluation of information and data about the performance of the process, will allow detection of undesired process variability. Evaluating the performance of the process identifies problems and determines whether action must be taken to correct, anticipate, and prevent problems so that the process remains in control.

An ongoing program to collect and analyze product and process data that relate to product quality must be established. The data collected should include relevant process trends and quality of incoming materials or components, in-process material, and finished products. The data should be statistically trended and reviewed by trained personnel. The information collected should verify that the quality attributes are being appropriately controlled throughout the process.

We recommend that a statistician or person with adequate training in statistical process control techniques develop the data collection plan and statistical methods and procedures used in measuring and evaluating process stability and process capability. Procedures should describe how trending and calculations are to be performed and should guard against overreaction to individual events as well as against failure to detect unintended process variability. Production data should be collected to evaluate process stability and capability. The quality unit should review this information. If properly carried out, these efforts can identify variability in the process and/or signal potential process improvements.

Good process design and development should anticipate significant sources of variability and establish appropriate detection, control, and/or mitigation strategies, as well as appropriate alert and action limits. However, a process is likely to encounter sources of variation that were not previously detected or to which the process was not previously exposed. Many tools and techniques, some statistical and others more qualitative, can be used to detect variation, characterize it, and determine the root cause. We recommend that the manufacturer use quantitative, statistical methods whenever appropriate and feasible.

We recommend continued monitoring and sampling of process parameters and quality attributes at the level established during the process qualification stage until sufficient data are available to generate significant variability estimates. These estimates can provide the basis for establishing levels and frequency of routine sampling and monitoring for the particular product and process. Monitoring can then be adjusted to a statistically appropriate and representative level. Process variability should be periodically assessed and monitoring adjusted accordingly.

Variation can also be detected by the timely assessment of defect complaints, out-of-specification findings, process deviation reports, process yield variations, batch records, incoming raw material records, and adverse event reports. Production line operators and quality unit staff should be encouraged to provide feedback on process performance. We recommend that the quality unit meet periodically with production staff to evaluate data, discuss possible trends or undesirable process variation, and coordinate any correction or follow-up actions by production.

Data gathered during this stage might suggest ways to improve and/or optimize the process by altering some aspect of the process or product, such as the operating conditions (ranges and set-points), process controls, component, or in-process material characteristics. A description of the planned change, a well-justified rationale for the change, an implementation plan, and quality unit approval before implementation must be documented. Depending on how the proposed change might affect product quality, additional process design and process qualification activities could be warranted.

Maintenance of the facility, utilities, and equipment is another important aspect of ensuring that a process remains in control. Once established, qualification status must be maintained through routine monitoring, maintenance, and calibration procedures and schedules. The equipment and facility qualification data should be assessed periodically to determine whether re-qualification should be performed and the extent of that re-qualification. Maintenance and calibration frequency should be adjusted based on feedback from these activities.

参考译文

下面的部分，描述了工艺验证总则、建议的工艺验证阶段和产品生命周期内每一阶段的特殊活动。

1. 工艺验证总则

在产品生命周期的所有阶段，科学良好的项目管理和归档将使得工艺验证项目更为有效和更具效率。下述规范应保证工艺相关信息的统一收集和评价，并确保提升这些信息在产品生命周期中易得性。

（1）我们建议由一个具有多学科（如工艺过程、工业药学、分析化学、微生物学、统计学、工业制造和质量保证）专业知识的综合团队进行工艺验证。项目计划，以及高级管理团队的全力支持，是成功进行验证的基本要素。

（2）在整个产品生命周期，可启动不同的研究，以发现、观察、关联或确认有关产品和工艺的信息。所有的研究应根据可靠的科学原则来计划和进行，妥善记录，并按照适用于生命周期相应阶段的既定程序予以批准。

（3）本指南中的术语属性（如质量、产品、组分）和参数（如工艺、操作和设备）不以其关键程度加以分类。在整个证明周期中，使用生命周期法进行基于风险决策的工艺验证时，将重要程度视作连续态比视作二元态更为有用所有的属性和参数，应该从它们在工艺中发挥的作用和对产品或在制品的影响的角度进行评估，并在获取新信息时重新进行评估。对这些属性或参数的控制程度，应该与其对工艺和工艺输出的风险相称。即，对风险较高的属性或参数，适宜采取更高程度的控制水平。本机构了解到术语的使用可能有差异，并希望所有生产商就其术语的含义和内涵及术语的分类方法与本机构进行沟通。

（4）很多产品是单一来源或涉及复杂的生产工艺。一个批次内的均一性和批间一致性是工艺验证活动的目标。验证为合理保护工艺，防止引入可能影响产品产量、引起供应问题及对公共健康产生负面影响的多种来源的产品提供了保证。

2. 第一阶段——工艺设计

工艺设计是明确商品化制造工艺的活动，该工艺在计划的主生产和控制记录中体现。本阶段的目的在于，设计一套适合常规商业化生产的工艺，使之能够持续地产出符合其质量属性的产品。

（1）认识和理解工艺　通常，早期工艺设计实验不需在CGMP（国际GMP规范）条件下进行，这种条件是用于商业流通的药品在第二阶段（工艺确认）和第三阶段（持续工艺验证）所必需。但它们应该按照包括文件编制管理规范在内的可靠的科学方法和原则进行。该建议与ICH Q10《药品质量体系》一致。管理的决策和正当理由应有足够文件证明并经内部审核，以确认和维护它们在随后的工艺和产品生命周期内的应用和变更价值。

尽管经常在小型实验室中开展，绝大部分的病毒灭活和杂质清除研究不能被视为早期工艺设计实验。在工业生产中为评价和判断产品质量而进行的病毒和杂质清除研究应该具备质量部门监督水准，这样的监督水准能够保证研究遵照可靠的科学方法和原则，并保证结论由数据支持。

产品开发活动为工艺设计阶段提供重要的基础资料，如打算采用的剂型、质量属性和一般生产途径。从产品开发活动中获得的工艺信息可在工艺设计阶段应用。工艺设计中，商品化制造设备的设计功能和局限应予以考虑，对生产过程中的不同的成分批次、生产操作人员、环境条件和测量系统可能引起的变化的预期影响也应予以考虑。然而，在此阶段，商品化生产中典型因素的变化尚不完全知晓。设计用来代表商品化工艺的实验室或中试规模的模型可以用来估量变化。

运用有效的工艺控制方法设计有效工艺，有赖于对工艺的认识和理解。通过揭示关系，包括变量输入（如组分特性或工艺参数）和结果输出（如在制品、中间体或成品）之间的多元相互作用，实验设计研究能够帮助促进工艺认识。风险分析工具可用于筛选实验设计研究的潜在变量以最大限度地减少开展的实验总数，同时使获得的认知最大化。实验设计研究的结果能为建立待用组分质量、设备参数和在制品质量属性范围提供依据。FDA（食品和药品管理局）一般不希望生产商直至工艺失败，还在对它进行开发和检测。

其他活动，如实验室或中试规模的实验或演示，也有助于某些条件评估和商品化工艺性能预测。这些活动也提供了能够用于模拟或模仿商品化工艺的信息。对某些单元操作或动态的计算机模拟或虚拟仿真能够促进对工艺的认知，并帮助避免工业规模生产中的问题。了解模型对工业规模工艺的体现程度（包括可能存在的任何差异）是重要的，原因在于这可能对源于模型的信息相关性有影响。

至关重要的是，用文件记录促使工艺理解的活动和研究。文件记录应反映工艺的决策基础。例如，生产商应该用文件记录单元操作研究的可变因素，以及这些变化因素被认为是有意义的理由。该信息在工艺确认和持续工艺验证阶段是有用的，包括设计修改或控制策略完善或变更的情况。

(2) 建立工艺控制策略 工艺知识及对它的理解是对所有单元操作和工艺总体上建立工艺控制方法的基础。工艺控制策略可设计用来减少输入变化，在生产中调整输入变化（并因此降低对产出的影响），或将两种方法结合。

工艺控制强调变化以保证产品质量。控制可由重要工艺控制点的物料分析与设备监控组成。与工艺控制类型和范围有关的决策可以借助于更早时候开展的风险评估，之后可随工艺经验的获得以加强和改进。

FDA 期望的控制措施包括物料质量检查和设备监控。对贯穿工作极限和生产过程监控的工艺控制，在两种可能的情况下，有必要特别关注：

① 由于取样或可检测性限制（例如病毒清除或微生物污染），产品属性不容易测量的情况，或

② 中间体和产物不能被高度表征，以及定义明确的质量属性不能被确认的情况。

这些控制措施是在主生产和控制记录中建立的。

更为先进的策略，可能涉及过程分析技术（PAT）的应用，可能包括及时分

析和控制回路以适应工艺条件,使产出保持恒定。这种类型的生产体系能比非过程分析技术体系提供更高程度的工艺控制。在使用 PAT 策略的情况下,工艺确认方法将有异于用于其他工艺设计中的方法。

计划中的商品化生产和控制记录,当中包含工艺控制的工作极限和总体策略,应转入下一阶段进行确认。

3. 第二阶段——工艺确认

在工艺验证的工艺确认(PQ)阶段,将对工艺设计进行评价以确认是否能够使用该工艺进行可重复的工业生产。该阶段具有两个因素:① 厂房设施设计以及设备和公用设施确认,以及② 工艺性能确认(PPQ)。在第二阶段,必须遵照符合 CGMP 的程序。进入商业流通前,必须成功完成第二阶段。在此阶段生产的产品,如果合格,可以放行流通。

(1) 厂房设施设计以及公用设施与设备确认　根据 CGMP 法规 211 部分 C 节对建筑和厂房设施的要求,生产厂房设施需要正确设计。至关重要的是,为保证正确的厂房设施设计和试运行而进行的活动应先于 PPQ。在这里,术语确认指为证明公用设施及设备符合预期用途和运转正确而从事的活动。这些活动必需先于工业规模的产品生产。

公用设施和设备确认一般包括下述活动。

① 基于公用设施和设备的特定用途选择工程物资、操作原则以及性能特点。

② 确认公用设施体系和设备遵照设计规范建造与安装(例如,使用适当材料、产能和功能按照设计建造,并正确连接和校准)。

③ 确认在全部预期的运行范围内,公共设施系统和设备运行与工艺要求相一致。该确认应包括在日常生产预期同等负荷下考察设备或系统功能。该确认还应包括考察预期的常规生产条件下的干预、停止和启动性能。运行范围应能够满足日常生产过程中的需要。

公用设施与设备确认可以有独立的计划,也可作为一项整体项目计划的部分。计划应考虑使用需要,并融入风险管理,使某些活动得以优先进行,并从确认活动中的运行和文件两方面确定投入水平。计划应确定以下事项:

要使用的研究或检测;

适用于评价结果的标准;

确认活动的时机;

相关部门和质量部门的责任;

文件编制和批准确认的程序。

项目计划还应包括公司对变更评价的要求。确认活动应用文件记录,并在报告中用含有强调计划标准的结论加以总结。质量控制部门必须审核和批准确认计划和报告。

(2) 工艺性能确认　工艺性能确认（PPQ）是第二阶段工艺确认的第二个要素。PPQ 将实际设施、公用系统、设备（已分别确认）以及训练有素的人员，与商品化生产工艺、控制程序以及各种生产要素结合在一起来生产商业产品。成功的 PPQ 将对工艺设计给予确认，并证明商品化制造工艺与预期一样。

此阶段的成功，在产品生命周期中标志着一个重要的里程碑。生产商着手药品商业流通之前，必须成功完成 PPQ。开始商业流通的决策应有来自于商品化规模生产批次的数据支持。来自于实验室小试和中试研究的数据能为商品化生产工艺表现达到预期提供额外的保障。

PPQ 方法应该以可靠的科学以及生产商对产品和工艺的理解和可论证的控制措施的总体水平为基础。来自所有相关研究的累积数据（如经过设计的实验、实验室小试、中试以及商业化批次）应用于建立 PPQ 中的生产条件。为充分理解商品化工艺，生产商需考虑规模效应。不过，如果有工艺设计数据提供保证，通常不需要在商品化大规模生产探索整个运行范围。类似的产品和工艺在之前可信的经验也有帮助。此外，只要可行并有意义，我们强烈建议公司使用客观测量的方法（如统计度量）获得足够的保证。

在绝大多数情况下，与典型的常规商品化生产相比，PPQ 具有较高的取样和附加检测水平，以及更仔细的工艺性能详查。监测和检验水平应足以在整个批次内确保产品质量始终如一。在适当的情况下，审查、检测和取样程度增加应持续贯穿工艺验证阶段，以建立特定产品和工艺常规取样和监测水平和频率。加强取样和监测周期的持续时间要考虑的问题可能包括，但不局限于产量、工艺复杂性、工艺理解水平和类似产品及工艺的经验。

某些物料，例如柱填充树脂或分子过滤介质，在对产品质量没有不利影响的情况下可以再利用，再利用程度可在相关实验室研究中进行评价。这种材料的可用寿命应在商品化生产中通过正在进行的 PPQ 确定。

使用 PAT 的生产工艺可能证明了一种不同的 PPQ 方法。PAT 工艺被设计用来实时检测一种在制品的多种属性，并在随后通过实时控制环路对工艺进行调整，以使工艺保持产出物料具有预期的质量。工艺设计阶段和工艺确认阶段应以待检测属性的检测系统和控制环路为中心。无论如何，验证任何生产工艺的目的相同的：为工艺可重现并能始终生产优质产品建立科学证据。

(3) PPQ 方案　规定了生产条件、控制、检测和预期结果的书面方案对工艺验证的这一过程至关重要。我们建议方案讨论下述要素。

① 生产条件，包括运行参数、工艺极限和组分（原材料）投入。
② 待收集数据以及何时、如何对其进行评估。
③ 每一重要工艺步骤需开展的检测（过程、放行、鉴定）以及可接受标准。
④ 取样方案，包括每一单元操作及属性的取样点、样品数和取样频率。样品数应该对批内和批间质量均足以提供统计学置信度。选定的置信水平可以建立

在考察中的特殊属性相关的风险分析的基础上。此阶段取样应比日常生产中的典型取样更广泛。

⑤ 考虑到基于科学和风险决策的有关持续生产合格品能力的标准和工艺性能指标。这些标准应包括：

用于分析所有收集数据的统计学方法描述（如定义批内及批间变化性的统计度量）；

用于处理预期条偏差件和处理非一致性数据的规定。就 PPQ 而言，如果没有文件证明和基于科学的正当理由，数据不应被排除于进一步的考虑之外。

⑥ 之前未完成的厂房设施设计和公用系统及设备确认、人员培训与确认以及物料来源的确认（组分和容器/密闭材料）。

⑦ 用于工艺、在制品和产品测定的分析方法的验证状态。

⑧ 相应部门及质量部门对方案的审核和批准。

(4) PPQ 方案执行与报告　在质量部门在内的相关部门对方案审核并做出批准前，不应开始执行 PPQ 方案。对方案的任何偏离，必须按照方案中既定的程序或规定进行处理。这种偏离在实施前必须由所有相应部门和质量部门证明合理并批准。

在 PPQ 方案执行期间，必须遵照商品化制造工艺和日常程序。PPQ 批次应在正常条件下由工艺中要求的日常生产中每一单元操作中的每一步骤的人员生产。正常操作条件应包括公用设施系统（如空气处理和水纯化）、物料、人员、环境和生产工序。

PPQ 方案执行完成之后，应遵循方案及时进行报告的编写及评价工作。该报告应：

① 讨论并相互参照方案的所有方面；

② 按照方案规定，总结并分析所收集的数据；

③ 对任何异常和方案中没有规定的额外数据进行评估；

④ 总结和讨论所有生产中的不符合项，例如偏差、异常检测结果或与工艺有效性有关的其他信息；

⑤ 充分详细地说明应该对现行程序与控制措施采取的所有整改措施或变更；

⑥ 对数据是否证明工艺符合方案建立的条件，和工艺是否被认为处于受控状态，详述明确结论。如果不是，该报告应该声明，得到这样一个结论之前还应该做什么。对于批准工艺和按照该工艺生产、放行并进入市场的批次，这个结论应基于文件证明的正当理由，并应考虑到从设计阶段到工艺确认阶段获得的所有知识和信息总和；

⑦ 包括所有相应部门和质量部门审核和批准。

4. 第三阶段——持续工艺验证

第三个验证阶段的目标，是在商品化生产期间持续保证工艺处于受控状态

(已验证状态)。用于检测计划外偏离设计工艺的系统对完成这一目标至关重要。遵守 CGMP 要求，特别是，收集和评估关于工艺性能的信息和数据，使检测出预期外的工艺变化成为可能。评估工艺性能来确定问题和决定是否必须采取措施纠正、预见和预防问题来确保工艺处于受控状态。

必须建立一个持续的程序，收集和分析与产品质量有关的产品和工艺数据。所搜集的数据，应包括相关的工艺趋势和入厂物料或组分、在制品和成品质量。数据应进行统计学趋势分析，并由经过培训的人员审核。对所收集信息，应核实质量属性在整个工艺中正受到适当控制。

我们建议，由统计学家或是在统计过程控制技术方面受过充分训练的人员，开发用于测定和评估工艺稳定性和工艺能力的数据收集方案、统计学方法及程序。程序应说明如何进行趋势分析和计算，还应防止对个别事件的过度反应，以及防止不能探测到意外的工艺变化。应收集生产数据对工艺稳定性和工艺能力进行评估。质量部门应审核这类信息。如果正确实施，这些努力能够甄别出工艺变异和/或潜在的工艺改进的趋势。

良好的工艺设计和开发，应能提前预见变异的重要来源，建立适当的探测、控制和/或减轻策略，以及适当的警报和行动限制。然而，一项工艺可能会遇到之前没有探测到，或工艺之前没有显示出的多个变异来源。很多工具和技术（一些属于统计学的，更多属于定性的）能用来探测变异，表征其特征，以及确定根本原因。我们建议生产商在任何适当和可行的情况下，使用定量的、统计学的方法。

我们建议，在工艺确认阶段已经建立的水平上对工艺参数和质量属性进行持续监测和取样，直到获得足够的数据用以进行显著性变异评估。这些评估能为建立特定产品和工艺的日常取样及监测的水平和频率提供基础。监测能因此调整到一个统计学上适当的、具有代表性的水平。工艺变异应定期进行评价，并相应地对监测做出调整。

通过及时评价缺陷投诉、对不合规格的调查结果、工艺偏离报告、工艺产率差异、批报告、入厂原料报告以及不良事件报告等，可以探测到变异。应鼓励生产线操作人员和质量部门员工提供对工艺性能的反馈。我们建议质量部门定期与生产部门人员开会，评估数据、讨论预期外的工艺变异，并通过生产协调所有整改或后续行动。

本阶段收集到的数据，可能为通过改变工艺或产品的一些方面，如操作条件（范围或设置点）、工艺控制、组分或在制品特性对工艺进行改进和/或优化。计划中的变更描述、变更的合理理由、实施计划以及实施前质量部门的批准，必须以文件记录。根据提议的变更可能对产品质量的影响，可能需要额外的工艺设计和工艺确认活动。

厂房设施、公用系统和设备的维护是确保工艺保持受控状态的另一个重要方面。一旦建立，该确认状态必须通过例行监测、维护与校验程序和时间表来保持。设备与厂房设施的确认数据应进行定期评价，以确定是否应开展再确认及再确认的范围。维护和校验频率应基于这些活动的反馈予以调整。

Further Reading

Introduction to the GMP Guide

The pharmaceutical industry of the European Union maintains high standards of Quality Management in the development, manufacture and control of medicinal products. A system of marketing **authorisations** ensures that all medicinal products are assessed by a competent authority to ensure compliance with contemporary requirements of safety, quality and efficacy. A system of manufacturing authorisations ensures that all products authorised on the European market are manufactured/imported only by authorised manufacturers, whose activities are regularly inspected by the competent authorities, using Quality Risk Management **principles**. Manufacturing authorisations are required by all pharmaceutical manufacturers in the European Union whether the products are sold within or outside of the Union.

Two **directives** laying down principles and **guidelines** of good manufacturing practice (GMP) for medicinal products were adopted by the Commission. Directive 2003/94/EC applies to medicinal products for human use and Directive 91/412/EEC for **veterinary** use. Detailed guidelines in accordance with those principles are published in the Guide to Good Manufacturing Practice which will be used in assessing applications for manufacturing authorizations and as a basis for **inspection** of manufacturers of medicinal products.

All Member States and the industry agreed that the GMP requirements applicable to the manufacture of veterinary medicinal products are the same as those applicable to the manufacture of medicinal products for human use. Certain detailed adjustments to the GMP guidelines are set out in two **annexes** specific to veterinary medicinal products and to **immunological** veterinary medicinal products.

The Guide is presented in three parts and supplemented by a series of annexes. Part I covers GMP principles for the manufacture of medicinal products. Part II covers GMP for active substances used as starting materials. Part III contains GMP related documents, which clarify regulatory expectations.

In addition to the general matters of Good Manufacturing Practice outlined in Part I and II, a series of annexes providing detail about specific areas of activity is included. For some manufacturing processes, different annexes will apply simultaneously (e.g. annex on **sterile preparations** and on **radiopharmaceuticals** and/or on biological medicinal products).

A **glossary** of some terms used in the Guide has been incorporated after the annexes. Part III is intended to host a collection of GMP related documents, which are not detailed guidelines on the principles of GMP laid down in Directives 2003/94/EC and 91/412/EC. The aim of Part III is to clarify regulatory expectations and it should be viewed as a source of information on current best practices. Details on the applicability will be described separately in each document.

The Guide is not intended to cover security aspects for the personnel engaged in manufacture. This may be particularly important in the manufacture of certain medicinal products such as highly active, biological and radioactive medicinal products. However, those aspects are governed by other provisions of Community or national law.

Throughout the Guide it is assumed that the requirements of the Marketing Authorisation relating to the safety, quality and efficacy of the products, are systematically incorporated into all the manufacturing, control and release for sale arrangements of the holder of the Manufacturing Authorisation.

The manufacture of medicinal products has for many years taken place in accordance with guidelines for Good Manufacturing Practice and the manufacture of medicinal products is not governed by **CEN/ISO** standards. Harmonised standards as adopted by the European standardisation organisations CEN/ISO may be used at industry's discretion as a tool for implementing a quality system in the pharmaceutical sector. The CEN/ISO standards have been considered but the terminology of these standards has not been implemented in this edition. It is recognised that there are acceptable methods, other than those described in the Guide, which are capable of achieving the principles of Quality Assurance. The Guide is not intended to place any restraint upon the development of any new concepts or new technologies which have been validated and which provide a level of Quality Assurance at least equivalent to those set out in this Guide.

New Words

authorisation [ˌɔːθərɪ'zeɪʃən] n. 授权，批准；批准（或授权）的证书
principle ['prɪnsəpl] n. 原则，原理；准则，道义；道德标准；本能
directive [dɪ'rektɪv] n. 指示；指令 adj. 指导的；管理的
guideline ['gaɪdˌlin] n. 指导方针；参考
veterinary ['vetərɪnərɪ] adj. 兽医的 n. 兽医
inspection [ɪn'spekʃn] n. 视察，检查
annex [ə'neks] n. 附加物；附属建筑物 vt. 附加；获得；并吞
immunological [ˌɪmjʊnə'lɒdʒɪkəl] adj. 免疫学的

sterile preparation ['sterəl͵prepə'reiʃən] n. 无菌制剂
radiopharmaceutical ['reidiəu͵fɑːmə'sjuːtikəl] n. 放射药剂；（供诊断或治疗用的）放射性药物 adj. 放射药剂的；（供诊断或治疗用的）放射性药物的
glossary ['glɒs(ə)rɪ] n. 术语（特殊用语）表；词汇表；专业词典
CEN abbr. 欧洲标准化委员会（The European Committee for Standardization）
ISO abbr. 国际标准化组织（International Standardization Organization）

参考文献

[1] FDA. Guidance for Industry-Process Validation: General Principles and Practices.2011.
[2] 何国强. 欧盟 GMP/GDP 法规汇编（中英文对照版）. 北京：化学工业出版社，2014.

Chapter 2 Non-Penicillin β-Lactam Drugs: A CGMP Framework for Preventing Cross-Contamination 预防非青霉素 β- 内酰胺类药物交叉污染的 CGMP 框架

1. Regulatory framework

Section 501(a) (2) (B) of the Federal Food, Drug, and Cosmetic Act [21 U.S.C. 351(a)(2)(B)] requires that, with few exceptions, all drugs be manufactured in compliance with current good manufacturing practices (CGMPs). Drugs that are not in compliance with CGMPs are considered to be **adulterated**. Furthermore, finished pharmaceuticals are required to comply with the CGMP regulations at 21 CFR parts 210 and 211.

Several CGMP regulations directly address facility and equipment controls and cleaning. For example, §211.42(c) requires building and facility controls in general to prevent cross-contamination of drug products. Specifically, the regulation states, "there shall be separate or defined areas or such other control systems for the firm's operations as are necessary to prevent contamination or **mix-ups**" during manufacturing, processing, packaging, storage, and holding.

With respect to penicillin, §211.42(d) requires that "operations relating to the manufacture, processing, and packing of penicillin shall be performed in facilities separate from those used for other drug products for human use." However, FDA has clarified that separate buildings may not be necessary, provided that the section of the manufacturing facility dedicated to manufacturing penicillin is isolated (i.e., completely and comprehensively separated) from the areas of the

New Words and Expressions

adulterated [əˈdʌltəreɪtɪd]
adj. 掺入次级品的

mix-up [ˈmɪksʌp]
n. 混战；混淆

facility in which non-penicillin products are manufactured. Under §211.46(d), manufacturers must completely separate air handling systems for penicillin from those used for other drugs for human use. Additionally, §211.176 requires manufacturers to test non-penicillin drug products for penicillin where the possibility of exposure to cross-contamination exists, and prohibits manufacturers from marketing such products if detectable levels of penicillin are found.

Although FDA has not issued CGMP regulations specific to **APIs**, the Agency has provided guidance to API manufacturers in the guidance for industry, **ICH** Q7, Good Manufacturing Practice Guidance for Active Pharmaceutical Ingredients. Because some APIs are sensitizing compounds that may cause **anaphylactic** shock, preventing cross-contamination in APIs is as important as preventing cross-contamination in finished products. The ICH Q7 guidance recommends using dedicated production areas, which can include facilities, air handling equipment and processing equipment, in the production of highly sensitizing materials, such as penicillins and cephalosporins.

2. *β*-**lactam** antibiotics

β-lactam antibiotics, including penicillins and the non-penicillin classes, share a basic chemical structure that includes a three-carbon, one-nitrogen cyclic amine structure known as the *β*-lactam ring. The side chain associated with the *β*-lactam ring is a variable group attached to the core structure by a peptide bond; the side chain variability contributes to antibacterial activity. As of the date of this publication, FDA has approved over 34 *β*-lactam

API (Active Pharmaceutical Ingredients 的简称)
n. 药物活性成分；原料药
ICH (International Conference on Harmonization 的简称)
n. 人用药品注册技术要求国际协调会议
anaphylactic [ˌænəfɪˈlæktɪk]
adj. 过敏的；过敏性的；导致过敏的

lactam [ˈlæktæm]
n. 内酰胺

compounds as active ingredients in drugs for human use. β-lactam antibiotics include the following five classes:
- Penicillins
- Cephalosporins
- **Penems**
- **Carbacephems**
- **Monobactams**

Allergic reactions associated with penicillins and non-penicillin β-lactams range from rashes to life-threatening anaphylaxis. **Immunoglobulin** E (IgE) antibodies mediate the immediate **hypersensitivity** reactions that are responsible for the symptoms of hay fever, asthma, **hives**, and **anaphylactic shock**. IgE-mediated hypersensitivity reactions are of primary concern because they may be associated with significant morbidity and mortality. There is evidence that patients with a history of hypersensitivity to penicillin may also experience IgE-mediated reactions to other β-lactams, such as cephalosporins and penems.

All non-penicillin β-lactams also have the potential to sensitize individuals, and subsequent exposure to penicillin may result in severe allergic reactions in some patients. Although the frequency of hypersensitivity reactions due to cross-reactivity between β-lactam classes can be lower than the risk within a class, the hazard posed is present and potentially life-threatening. The potential health hazard of non-penicillin β-lactams therefore is similar to that of penicillins. Further similarities between non-penicillin β-lactams and penicillins are as follows:
- It is difficult to define the minimal dose below which allergic responses are unlikely to occur in humans.

penem['penəm]
n. 青霉烯类
carbacephem [kɑː'beɪkɪphəm]
n. 碳头孢烯类
monobactam [mɒ'nɒbæktəm]
n. 单环菌素
immunoglobulin [ˌɪmjʊnəʊ'glɒbjʊlɪn]
n. 免疫球蛋白；免疫血球素
hypersensitivity [ˌhaɪpəˌsensɪ'tɪvɪtɪ]
n. 过敏症；高灵敏度
hives [haɪvz]
n. 荨麻疹；假膜性喉头炎
anaphylactic shock [ˌænəfɪ'læktɪk ʃɔk]
n. 过敏性休克；过敏反应休克

- There is a lack of suitable animal or receptor testing models that are predictive of human sensitivity.
- The **threshold** dose at which allergenic response could occur is extremely low and difficult to detect with current analytical methods.

While β-lactam antibiotics are similar to one another in many ways, they may differ in **pharmacokinetics**, antibacterial activity, and potential to cause serious allergic reactions. Because allergy testing methods have not been well-validated, it is clinically difficult to determine the occurrence and rate of cross-reactivity between β-lactam antibiotics in humans. Therefore, undiagnosed or underreported cases of cross-reactivity likely exist. Some β-lactam antibiotics have negligible potential for cross-reactivity with β-lactams of other classes, whereas other β-lactam compounds may exhibit sensitizing activity as derivatives before the incorporation of side chains that confer antibacterial activity.

Regardless of the rate of cross-reactivity between β-lactam drugs or the mechanism of action by which such cross-reactivity may occur, the potential health risk to patients indicates that drug manufacturers should take steps to control for the risk of cross-contamination for all β-lactam products.

3. β-lactamase inhibitors

β-lactam compounds such as clavulanic acid, **tazobactam**, and **sulbactam** have weak antibacterial activity but are irreversible inhibitors of many β-lactamases. These compounds, which are potential sensitizing agents, are typically used

threshold [ˈθreʃəʊld]
n. 极限；临界值

pharmacokinetics [ˌfɑːməkəʊkiˈnetɪks]
n. 药物（代谢）动力学

tazobactam [təzɒˈbæktəm]
n. 他唑巴坦

sulbactam [sʌlˈbæktəm]
n. 舒巴坦

in combination with specific β-lactam agents to preserve antibacterial activity (e.g., amoxicillin-clavulanate, **piperacillin**-tazobactam). Because these compounds are almost always used in combination with specific β-lactam agents, any clinical observations of hypersensitivity reactions likely would be attributed to the β-lactam antibiotic component rather than the inhibitor. Although there have been no case reports confirming anaphylactic reactions to a β-lactamase inhibitor that is also a β-lactam, these compounds are potentially sensitizing agents, and manufacturers should implement controls to reduce the risk of cross-contamination with β-lactamase inhibitors as with all other β-lactam products.

piperacillin [paɪpərəˈsɪlɪn] n. 哌拉西林

4. β-lactam intermediates and derivatives

Some β-lactam intermediate compounds and derivatives also possess similar sensitization and cross-reactivity properties. β-lactam intermediate compounds usually are API precursor materials that undergo molecular change or purification before use in the manufacture of β-lactam antibiotic APIs. As a result of these changes, the intermediate compounds may develop antigenic characteristics that can produce allergic reactions. For example, 6-aminopenicillanic acid (6-APA) serves as the intermediate for the formation of all synthetic penicillins that are formed by attaching various side chains. The structure of 6-APA includes unbroken β-lactam and thiazolidine rings. The β-lactam ring is relatively unstable, and it commonly breaks open. In the case of 6-APA, this breakage leads to the formation of a penicilloyl moiety, which is the major antigenic determinant of penicillin. This moiety is thought to be a common cause of penicillin urticarial reaction.17 Degradation of 6-APA can also result in the formation of minor antigenic determinants, including penicilloic acids, penaldic acid, and penicillamine. Anaphylactic reactions to penicillins usually are due to the presence of IgE antibodies to minor determinants in the body. Although 6-APA is not a true antibiotic, it still carries with it a potential to induce allergenicity.

Derivatives are unintended by-products that occur during the manufacturing process (i.e., an impurity or degradant). Like intermediates, β-lactam derivatives could have sensitizing properties and may develop antigenic properties that can produce allergic reactions. β-lactam chemical manufacturing processes including, but not limited to, fermentation and synthesis, may create β-lactam intermediates or derivatives with unknown health consequences. Although the health risk of sensitization and cross-reaction is difficult to predetermine for β-lactam intermediates and derivatives and is not always well-defined, manufacturing controls intended to reduce the risk of cross-contamination should be considered for operations that produce β-lactam intermediates or derivatives.

参考译文

1. 法规框架

联邦食品、药品和化妆品法案 501(a)(2)(B) 部分 21 U.S.C. 351(a)(2)(B) 要求：除少数例外情况，所有的药品都应在符合现行药品生产质量管理规范（cGMPs）的条件下生产。不按照 cGMP 条件生产的药物将被视为假药。此外，制剂生产还应遵守 21CFR210 和 211 部分 cGMP 法规要求。

一些 cGMP 法规对设施和设备的控制和清洁有直接的规定。例如，§211.42(c) 部分要求对建筑和设施进行控制以预防药物的交叉污染。这个规定特别指出：在药品生产、加工、包装、储存和处理过程中，"公司的操作应有独立或指定区域，或其他类似控制系统来预防药物污染和混淆"。

关于青霉素，§211.42（d）部分要求"进行有关青霉素生产、加工和包装的设施应与其他人用药的设施分开"。但是，FDA 也有说明，如果将专用于生产青霉素的生产设施与青霉素外的产品生产区域隔离（如完全和全面地隔开），并非一定要采用单独的建筑。§211.46（d）部分规定，药品生产商对青霉素生产必须采用单独的空气处理系统，不得和其他人用药物生产公用。另外，§211.176 部分要求药品生产商在青霉素可能暴露导致交叉污染时，检测非青霉素药物中的青霉素含量，并禁止销售可检测出青霉素的药物。

尽管 FDA 还没有发布专门针对原料药的 cGMP 法规，但当局为 API 生产商提供了行业指南 ICH Q7，即原料药的质量生产管理规范。由于一些原料药是可能导致过敏性休克的致敏化合物，因此，预防原料药的交叉污染和预防制剂的交叉污染同样重要。ICH Q7 指南推荐在生产青霉素和头孢菌素等高致敏性物料时采用专用的生产区域，包括专用的生产场地、空气处理设备和加工设备。

2. β- 内酰胺类抗生素

β- 内酰胺类抗生素，包括青霉素类和非青霉素类，具有共同的基础化学结构，包括一个含有 3 个 C 和 1 个 N 的环胺结构，即 β- 内酰胺环。与 β- 内酰胺相关的侧链为可变基团，通过肽键和核心结构相连，侧链的可变性赋予了其抗菌活性。截止至本指南发布之日，FDA 已经批准了超过 34 种 β- 内酰胺类化合物作为人用药物的活性成分。β- 内酰胺类抗生素包括以下五类：

- 青霉素类
- 头孢菌素类
- 青霉烯类
- 碳头孢烯类
- 单环菌素类

青霉素类和非青霉素 β- 内酰胺类药物可能引发包括从皮疹到危及生命的过敏症等不同程度的过敏反应。由免疫球蛋白 E（IgE）介导的超敏反应相关症状主要包括花粉症、哮喘、荨麻疹以及过敏性休克。IgE 介导的超敏反应是最需要关注的，因为它可能与高的发病率和死亡率有关。有证据证明有青霉素过敏史的患者在使用其他 β- 内酰胺类药物时可能也会发生这类 IgE 介导的超敏反应，如头孢菌素类和青霉烯类。

所有非青霉素 β- 内酰胺类药物也有可能导致部分个体敏化，当再次暴露于青霉素时，部分患者可能会出现严重的过敏反应。尽管不同类 β- 内酰胺类药物发生交叉反应导致的超敏反应频率比同类 β- 内酰胺类药物的低，但是产生的危害仍然存在并且可能危及生命。这种非青霉素 β- 内酰胺类药物的潜在风险和青霉素类抗生素相似，两者的相似之处还包括：

- 难以确定人不发生过敏反应的最低剂量
- 缺乏能预测人体过敏反应的合适的动物或受体试验模型
- 发生过敏反应的阈剂量极低,按现有分析方法难以进行检测。

尽管 β-内酰胺类抗生素在很多方面彼此都很相似,但在药代动力学、抗菌行为和导致严重过敏反应的可能性上可能有所分别。由于目前过敏检测的方法都缺乏充分的验证,在临床上很难对 β-内酰胺类抗生素之间是否会发生人体交叉反应及其发生的概率进行检测。因此,可能会存在临床上未诊断出或未汇报的交叉反应情形。一些 β-内酰胺类抗生素和其他类别的 β-内酰胺类抗生素的交叉反应风险可以忽略不计,但是别的 β-内酰胺类化合物和引入决定抗菌活性的侧链前的衍生物一样可能表现出致敏性能。

不管 β-内酰胺类药物之间发生交叉反应几率有多大或发生类似交叉反应的作用机制如何,都会对病人带来潜在的健康风险,因此,药品生产商应采取措施控制 β-内酰胺类产品之间的交叉污染风险。

3. β-内酰胺酶抑制剂

一些 β-内酰胺类化合物,如克拉维酸、他唑巴坦和舒巴坦虽然抗菌活性很弱,但却是很多 β-内酰胺酶的不可逆性抑制剂。这些化合物是潜在的致敏剂,通常和其他特定的 β-内酰胺类制剂联合使用,以维持抗菌活性(如阿莫西林-克拉维酸、哌拉西林-他唑巴坦)。因为这些化合物几乎总是和特定的 β-内酰胺类制剂联合使用,所有临床上观察到的超敏反很可能都是由于 β-内酰胺类抗菌组分的作用,而不是这些抑制剂的作用。虽然没有确切的报告能证实同样具有 β-内酰胺的 β-内酰胺酶抑制剂会导致过敏反应,但这些组分仍然是潜在的致敏剂,和其他的 β-内酰胺类产品一样,生产商也应采取控制措施降低 β-内酰胺酶抑制剂交叉污染的风险。

4. β-内酰胺中间体和衍生物

一些 β-内酰胺类中间体和衍生物也可能有类似的致敏性和交叉反应性。β-内酰胺类中间体一般是原料药的前体,经分子改变或纯化后用于 β-内酰胺类抗生素原料药的生产。这些改变可能使这些中间体成分具有引发过敏反应的致敏性能。例如,6-氨基青霉烷酸(6-APA)是形成所有合成类青霉素药物的中间体,青霉素类药物就是由这种中间体和各种各样的侧链结合后生成的。6-APA 结构包括一个未被破坏的 β-内酰胺环和一个噻唑环,β-内酰胺环并不稳定,容易被打开。对于 6-APA,β-内酰胺环打开后可以形成青霉噻唑基,这是青霉素的主要抗原决定簇,被认为是青霉素引发荨麻症反应的主要原因。6-APA 的降解还可能形成一些次要抗原决定簇,包括青霉素裂解酸、青霉醛酸和青霉胺。青霉素类抗生素的过敏反应主要是针对次要抗原决定簇的体内 IgE 抗体介导的,因此,尽管 6-APA 并不是一种真正的抗生素,但是仍然有导致过敏反应的风险。

衍生物是生产过程中形成的非预期副产物（如杂质或降解物）。比如中间体，β-内酰胺类衍生物可能具有致敏性，以及形成导致过敏反应的抗原特性。β-内酰胺化合物生产工艺包括但不局限于发酵和合成，这些工艺都可能产生未知的影响健康的中间体或衍生物。尽管β-内酰胺类中间体和衍生物的致敏性和交叉反应的健康风险很难确定，也很难被很好地定义，但是应当考虑采取措施控制产生β-内酰胺中间体或衍生物的工艺步骤，从而降低交叉污染的风险。

Further Reading
Validation of Cleaning Processes

General Requirements

FDA expects firms to have written procedures (SOP's) detailing the cleaning processes used for various pieces of equipment. If firms have one cleaning process for cleaning between different batches of the same product and use a different process for cleaning between product changes, we expect the written procedures to address these different **scenarios**. Similarly, if firms have one process for removing water soluble **residues** and another process for non-water soluble residues, the written procedure should address both scenarios and make it clear when a given procedure is to be followed. Bulk pharmaceutical firms may decide to dedicate certain equipment for certain chemical manufacturing process steps that produce **tarry** or **gummy** residues that are difficult to remove from the equipment. Fluid bed dryer bags are another example of equipment that is difficult to clean and is often dedicated to a specific product. Any residues from the cleaning process itself (detergents, solvents, etc.) also have to be removed from the equipment.

FDA expects firms to have written general procedures on how cleaning processes will be validated.

FDA expects the general validation procedures to address who is responsible for performing and approving the validation study, the acceptance criteria, and when revalidation will be required.

FDA expects firms to prepare specific written validation protocols in advance for the studies to be performed on each manufacturing system or piece of equipment which should address such issues as sampling procedures, and analytical methods to be used including the sensitivity of those methods.

FDA expects firms to conduct the validation studies in accordance with the protocols and to document the results of studies.

FDA expects a final validation report which is approved by management and which states whether or not the cleaning process is valid. The data should support a

conclusion that residues have been reduced to an "acceptable level."
New Words

scenario [sɪ'nɑːrɪəʊ] *n.* 方案；情节；剧本

residue ['rezɪdjuː] *n.* 残渣；剩余；滤渣

tarry ['tɑːrɪ] *n.* 逗留 *adj.* 涂了焦油的

gummy ['gʌmɪ] *adj.* 黏性的，胶粘的；含树胶的

参考文献

[1] FDA. Guidance for Industry-Non-Penicillin Beta-Lactam Drugs:A cGMP Framework for Preventing Cross-Contamination.2013.

[2] FDA. Validation of Cleaning Processes.http://www.fda.gov/iceci/inspections/inspectionguiins/ucm074922.htm .2014-11-25.